# Progress in Mathematical Physics
Volume 28

*Editors-in-Chief*
Anne Boutet de Monvel, *Université Paris VII Denis Diderot*
Gerald Kaiser, *The Virginia Center for Signals and Waves*

*Editorial Board*
D. Bao, *University of Houston*
C. Berenstein, *University of Maryland, College Park*
P. Blanchard, *Universität Bielefeld*
A.S. Fokas, *Imperial College of Science, Technology and Medicine*
C. Tracy, *University of California, Davis*
H. van den Berg, *Wageningen University*

Elena Obolashvili

# Higher Order Partial Differential Equations in Clifford Analysis

*Effective Solutions to Problems*

Birkhäuser
Boston • Basel • Berlin

Elena Obolashvili
Georgian Academy of Sciences
A. Razmadze Mathematical Institute
Tbilisi 380093
Georgia

**Library of Congress Cataloging-in-Publication Data**

Obolashvili, E. I. (Elena Irodionovna)
 Higher order partial differential equations in Clifford analysis : effective solutions to problems / Elena Obolashvili.
    p. cm.– (Progress in mathematical physics ; 28)
 Includes bibliographical references and index.
 ISBN 0-8176-4286-2 (alk. paper)  — ISBN 3-7643-4286-2 (alk. paper)
  1. Differential equations, Parabolic. 2. Clifford algebras. I. Title. III. Progress in mathematical physics ; v. 28.

QA377.O26 2002
515'.353–dc21                                                                                              2002026005
                                                                                                                  CIP

---

AMS Subject Classifications: Primary: 73-XX, 76-XX, 58-XX; Secondary: 65-XX, 63-XX, 79-XX

---

Printed on acid-free paper.
©2003 Birkhäuser Boston               *Birkhäuser*

All rights reserved. This work may not be translated or copied in whole or in part without the written permission of the publisher (Birkhäuser Boston, c/o Springer-Verlag New York, Inc., 175 Fifth Avenue, New York, NY 10010, USA), except for brief excerpts in connection with reviews or scholarly analysis. Use in connection with any form of information storage and retrieval, electronic adaptation, computer software, or by similar or dissimilar methodology now known or hereafter developed is forbidden.
The use of general descriptive names, trade names, trademarks, etc., in this publication, even if the former are not especially identified, is not to be taken as a sign that such names, as understood by the Trade Marks and Merchandise Marks Act, may accordingly be used freely by anyone.

ISBN 0-8176-4286-2     SPIN  10863036
ISBN 3-7643-4286-2

Reformatted from the author's files by John Spiegelman, Abbington, PA.
Printed in the United States of America.

9 8 7 6 5 4 3 2 1

Birkhäuser   Boston • Basel • Berlin
*A member of BertelsmannSpringer Science+Business Media GmbH*

# Contents

**Preface**   vii

## I   Boundary Value Problems for Regular, Generalized Regular and Pluriregular Elliptic Equations   1

### I   Two-Dimensional Cases   3
    0   Introduction . . . . . . . . . . . . . . . . . . . . . . . . . . . .   3
    1   BVP for Holomorphic Functions . . . . . . . . . . . . . . . . .   18
    2   BVP for Generalized Holomorphic Functions . . . . . . . . . . .   27
    3   Mixed BVP for Generalized Holomorphic Functions . . . . . . . .   32
    4   BVP for Beltrami and Generalized Beltrami Equations . . . . . . .   38
    5   BVP for Pluriholomorphic, Plurigeneralized Holomorphic, Polyharmonic, Polymetaharmonic Functions and the PluriBeltrami Equation . . . . . . . . . . . . . . . . . . . . . . . .   44
    6   Nonlocal Problems for Biholomorphic Functions . . . . . . . . . .   54
    7   More BVP for Pluriholomorphic and Plurigeneralized Holomorphic Functions . . . . . . . . . . . . . . . . . . . . . . . . . . . . . .   64

### II   Multidimensional Cases   67
    0   Introduction . . . . . . . . . . . . . . . . . . . . . . . . . . . .   67
       0.1   Elements of Clifford analysis . . . . . . . . . . . . . . . .   67
       0.2   The basic $L$ theory of the FIT. . . . . . . . . . . . . . . .   69

|   | 1 | BVP for Regular and Generalized Regular Functions, and the Hobson Formula in Clifford Analysis . . . . . . . . . . . . . . . . . . . . . 76 |
|---|---|---|
|   | 2 | BVP for Pluriregular, Generalized Pluriregular, and Polyharmonic Functions and the Poly-Helmholtz Equation . . . . . . . . . . . . . . 104 |
|   | 3 | BVP for Beltrami, Generalized Beltrami, and Pluri-Beltrami Equations in Clifford Analysis . . . . . . . . . . . . . . . . . . . . . . . . . . . 110 |
|   |   | 3.1  Classification in multidimensional space . . . . . . . . . . . . 110 |
|   |   | 3.2  BVP . . . . . . . . . . . . . . . . . . . . . . . . . . . . . . . 114 |
|   | 4 | More Problems for Pluriregular and Plurigeneralized Regular Functions . . . . . . . . . . . . . . . . . . . . . . . . . . . . 118 |

## II  Initial Value Problems for Regular and Pluriregular, Hyperbolic and Parabolic Equations     123

### III  Hyperbolic and Plurihyperbolic Equations in Clifford Analysis     125

    0    Introduction . . . . . . . . . . . . . . . . . . . . . . . . . . . . . . . . 125

    1    IVP for Hyperbolic Systems (Maxwell and Dirac Equations) . . . . . 132

    2    B&IVP for Pluriregular and Plurigeneralized Regular Hyperbolic Systems, Polywave and Poly-Klein–Gordon Equations, Harmonic-Wave and Harmonic Klein–Gordon Equations . . . . . . . 137

    3    Pluri-Beltrami Hyperbolic Equations . . . . . . . . . . . . . . . . . . 146

### IV  Parabolic and Pluriparabolic Equations in Clifford Analysis     151

    0    Introduction . . . . . . . . . . . . . . . . . . . . . . . . . . . . . . . . 151

    1    IVP for Parabolic Systems in Clifford Analysis . . . . . . . . . . . . 156

    2    Pluriparabolic Systems and Polyheat Equations . . . . . . . . . . . . 159

    3    Parabolic Regular Equations of the Second Kind and IVP . . . . . . . 161

    4    Elliptic-Parabolic, Hyperbolic-Parabolic and Elliptic-Hyperbolic-Parabolic Equations . . . . . . . . . . . . . . . . 165

**Epilogue**     171

**References**     173

**Index**     177

# Preface

*The most important thing is to write equations in a beautiful form and their success in applications is ensured.*

Paul Dirac

The uniqueness and existence theorems for the solutions of boundary and initial value problems for systems of high-order partial differential equations (PDE) are sufficiently well known. In this book, the problems considered are those whose solutions can be represented in quadratures, i.e., in an effective form. Such problems have remarkable applications in mathematical physics, the mechanics of deformable bodies, electromagnetism, relativistic quantum mechanics, and some of their natural generalizations. Almost all such problems can be set in the context of Clifford analysis. Moreover, they can be obtained without applying any physical laws, a circumstance that gives rise to the idea that Clifford analysis itself can suggest generalizations of classical equations or new equations altogether that may have some physical content. For that reason, Clifford analysis represents one of the most remarkable fields in modern mathematics as well as in modern physics.

The aim of this book is to consider the solution of systems of elliptic, hyperbolic and parabolic equations. In the scientific literature on Clifford analysis, more attention has been paid to elliptic equations (see [BDS], [GS], [Ry]). There are only a few papers in which hyperbolic equations are considered ([Ri], [He]) and none that consider parabolic equations. It must be indicated that parabolic equations were constructed for the first time in our paper [Ob5] in which initial value problems were solved. Herein we are only interested in solving problems effectively. The uniqueness and existence of the solutions is easily determined from their effective representations.

Boundary and initial value problems for elliptic and hyperbolic equations as natural multidimensional analogues of the generalized Cauchy–Riemann and Beltrami equations were considered for the first time in [Ob2], [Ob3], [Ob6]. Herein boundary and initial value problems for plurigeneralized regular and pluri-Beltrami equations are considered for the first time, where pluriregular equations are related to the equations

$$\Delta^m u(x) = 0, \quad \left(\Delta - \frac{\partial^2}{\partial t^2}\right)^m u(x,t) = 0, \quad \left(\Delta - \frac{\partial}{\partial t}\right)^m u(x,t) = 0, \quad m \geq 1,$$

$$\Delta\left(\Delta - \frac{\partial^2}{\partial t^2}\right) u(x,t) = 0, \quad \Delta\left(\Delta - \frac{\partial}{\partial t}\right) u(x,t) = 0,$$

$$\left(\Delta - \frac{\partial^2}{\partial t^2}\right)\left(\Delta - \frac{\partial}{\partial \tau}\right) u(x,t,\tau) = 0,$$

$$\Delta\left(\Delta - \frac{\partial^2}{\partial t^2}\right)\left(\Delta - \frac{\partial}{\partial \tau}\right) u(x,t,\tau) = 0, \quad x(x_0,\ldots,x_{n-1}),$$

called the polyharmonic, polywave, polyheat, harmonic-wave, harmonic-heat, wave-heat and harmonic-wave-heat equations, respectively, and $\Delta$ is the Laplace operator.

This book consists of two parts that are divided into chapters which are in turn divided into sections. Each chapter begins with an introduction in which some definitions and theorems are presented. I have attempted to write this book for the general reader, but also for the specialist in PDE. It can be used by mathematicians and physicists who are interested in boundary and initial value problems.

I would like to thank heartily my Spanish colleague and friend Professor Maria Pilar–Guzman, former President of the Institute of Statistics of the Community of Madrid, for financial and moral support and for her excessive kindness.

I am much indebted to the typist group of A. Razmadze Mathematical Institute headed by Maia Kvinikadze for their careful preparation of my handwritten manuscript.

# Higher Order Partial Differential Equations in Clifford Analysis

*Effective Solutions to Problems*

# Part I

# Boundary Value Problems for Regular, Generalized Regular and Pluriregular Elliptic Equations

*I think the theory of complex variables excellent since Cauchy's integrals have strong power.*

Paul Dirac

# I
# Two-Dimensional Cases

## 0 Introduction

Let $D^+$ be a finite multiply connected domain in the complex plane $z = x+iy$ bounded by piecewise-smooth lines: $\Gamma_0$ is a closed line; $\Gamma_1, \ldots, \Gamma_n$ are inside $\Gamma_0$. The lines $\Gamma_k (k = 0, \ldots, n)$ are positively oriented; i.e., $D^+$ is situated on the left when one moves in the positive direction along $\Gamma_k$. $D^-$ denotes the domain situated outside $\Gamma$, where $\Gamma \equiv \cup_0^n \Gamma_k$. If $D^+$ is an infinite domain, then its boundary is $\Gamma \equiv \cup_1^n \Gamma_k$.

Let $\varphi(t)$ be an integrable function on $\Gamma$. As is well known, the Cauchy-type integral [Mu1]

$$\phi(z) = \frac{1}{2\pi i} \int_\Gamma \frac{\varphi(t)dt}{t-z}, \tag{1}$$

is an analytic function in $D^+$ and $D^-$ that vanishes at infinity. Suppose $\varphi(t)$ is a Hölder-continuous function, i.e., $\varphi(t) \in C^\alpha(\Gamma)$, $0 < \alpha < 1$. Then $\phi(z)$ has the boundary values

$$\phi^+(t) = \lim_{D^+ \ni z \to t} \phi(z), \quad \phi^-(t) = \lim_{D^- \ni z \to t} \phi(z).$$

For $t \in \Gamma$ one has the Plemelj–Sokhotzki formulae

$$\phi^+(t) = \frac{1}{2}\varphi(t) + \phi(t),$$
$$\phi^-(t) = -\frac{1}{2}\varphi(t) + \phi(t),$$
(2)

where $\phi(t)$ is understood as the Cauchy principal value of (1). Moreover, the following theorem (Plemelj–Privalov) holds: $\phi^+(t)$ and $\phi^-(t)$ are Hölder-continuous functions $C^\alpha$. The proof can be found in [Be].

Let $\Gamma$ be the real axis and
$$\varphi(t) = \varphi(\infty) + O\left(\frac{1}{|t|^\alpha}\right), \quad \alpha > 0.$$

Then the Cauchy-type integral exists for $z = t_0 \in \Gamma$ as a Cauchy principal value, i.e.,
$$\int_{-\infty}^{+\infty} \frac{\varphi(t)dt}{t - t_0} = \lim_{\substack{\varepsilon \to 0 \\ N \to \infty}} \left[ \int_{-N}^{t_0-\varepsilon} \frac{\varphi(t)dt}{t - t_0} + \int_{t_0+\varepsilon}^{N} \frac{\varphi(t)dt}{t - t_0} \right]$$

Let $W(z) = u(x, y) - iv(x, y)$ be a solution of a Cauchy–Riemann system, which can be written in complex form as
$$\frac{\partial W}{\partial \bar{z}} = 0, \quad 2\frac{\partial}{\partial \bar{z}} = \frac{\partial}{\partial x} + i\frac{\partial}{\partial y}, \quad 2\frac{\partial}{\partial z} = \frac{\partial}{\partial x} - i\frac{\partial}{\partial y}$$
(3)

Two linearly independent fundamental solutions of these equations
$$\overset{1}{W} = u_1 - iv_1 = 2\frac{\partial \ln |z|}{\partial z} = \frac{1}{z},$$
$$\overset{2}{W} = u_2 - iv_2 = 2i\frac{\partial \ln |z|}{\partial z} = \frac{i}{z},$$
(4)

are connected with the kernel of the Cauchy-type integral. Linear independence requires that $c_1 \overset{1}{w} + c_2 \overset{2}{w} = 0$ for real constants $c_1, c_2$ only if $c_1 = c_2 = 0$.

Now consider two real functions $P(x, y)$, $Q(x, y)$ that have first-order continuous derivatives in $D^+$ and are continuous in $D^+ \cup \Gamma$. Then one has the well-known Green's formula (see, e.g., [TS])
$$\int_{D^+} \left(\frac{\partial P}{\partial x} + \frac{\partial Q}{\partial y}\right) dxdy = \int_\Gamma (P \cos nx + Q \cos ny) ds,$$

where $n$ is an outward unit normal vector of $\Gamma$.

Let 1) $P = u$, $Q = v$ and 2) $P = v$, $Q = -u$. Then by Green's formula one can easily obtain

$$2 \int_D \frac{\partial W}{\partial \bar{z}} dx dy = -i \int_\Gamma W(t) dt, \tag{5}$$

by taking into consideration the equalities

$$\cos nx = \cos sy = \frac{dy}{ds},$$
$$\cos ny = -\cos sx = -\frac{dx}{ds},$$

where $s$ is tangent to $\Gamma$. The axes $x$, $y$ and $n$, $s$ are similarly oriented; i.e., both are either left or right oriented. Equation (5) is a remarkable formula by force of which one can prove classical theorems such as the Cauchy theorem and Cauchy integral formula.

Let $W$, $W_1$ be two regular holomorphic functions. If we replace $W$ in (5) with $WW_1$, then by force of (3) one has

$$2 \int_D \left( W_1 \frac{\partial W}{\partial \bar{z}} + W \frac{\partial W_1}{\partial \bar{z}} \right) dx dy = -i \int_\Gamma W W_1 dt = 0.$$

Let $W_1 = \frac{1}{t-z}$ and $D_\varepsilon$ be the domain bounded by $\Gamma$ and $|t - z| = \varepsilon$. If we take $t - z = \varepsilon e^{-\vartheta}$, $0 \le \vartheta \le 2\pi$, it is not difficult to obtain

$$\lim_{\varepsilon \to 0} -i \int_{|t-z|=\varepsilon} \frac{W(t) dt}{t - z} = 2\pi W(z).$$

Thus, we have

$$W(z) = \frac{1}{2\pi i} \int_\Gamma \frac{W(t) dt}{t - z} + \frac{1}{\pi} \int_D \frac{\frac{\partial W}{\partial \bar{\zeta}} d\xi d\eta}{\zeta - z}, \quad z \in D. \tag{$5_1$}$$

Applying the above formulae one can construct effective solutions of the following boundary value problems (BVP) for holomorphic functions:

1. Hilbert BVP (Conjugate problem). Let $W(z)$ be a piecewise holomorphic function with a jump line $\Gamma$; i.e. it is holomorphic in $D^+$ and $D^-$. Find $W(z)$ that vanishes at infinity and satisfies the condition

# I. Two-Dimensional Cases

$$W^+(t) = G(t)W^-(t) + g(t), \quad t \in \Gamma, \tag{6}$$

where $G(t) \neq 0$, $g(t)$ are Hölder-continuous functions. It is also possible to solve this problem when $G(t)$ has a finite number of first-kind discontinuity points. That case was first considered by N. Muskhelishvili [Mu1]. Some mathematicians call the Gilbert problem the Riemann problem (e.g., [Be]).

2. Riemann–Hilbert BVP. Find a holomorphic function $W(z)$ in $D^+$ that satisfies the condition

$$\text{Re}[\lambda(t)W^+(t)] = g(t), \quad t \in \Gamma, \tag{7}$$

where $\lambda(t) \neq 0$, $g(t)$ are Hölder-continuous functions. To solve these BVP see [Mu1].

3. Compound BVP. Let $D_k$ be a domain bounded by a closed line $\Gamma_k$. Find a piecewise holomorphic function $W(z)$ in $D^+ + \bigcup_1^n D_k$ that satisfies the conditions

$$\text{Re}[\lambda(t)W^+(t)] = g_1(t), \quad t \in \Gamma_0, \tag{8}$$
$$W^+(t) = G(t)W^-(t) + g_2(t), \quad t \in \Gamma_k \; (k = 1, \ldots, n), \tag{9}$$

where $\lambda(t) \neq 0$, $G(t) \neq 0$, $g_1, g_2$ are given Hölder-continuous functions.

Consider some partial cases in which the solutions of the above problems can be written in a very simple way.

**Problem 1.** The solution of (6) for $G(t) = 1$, by force of (2), is

$$W(z) = \frac{1}{2\pi i} \int_\Gamma \frac{g(t)dt}{t-z}, \quad z \in D^+, D^-, \tag{10}$$

and if $G(t) = -1$

$$W(z) = \frac{1}{2\pi i} \int_\Gamma \frac{g(t)dt}{t-z}, \quad \text{for} \; z \in D^+,$$

$$W(z) = -\frac{1}{2\pi i} \int_\Gamma \frac{g(t)dt}{t-z}, \quad \text{for} \; z \in D^-. \tag{11}$$

The problem (7), (8), (9) can be solved effectively in a simple way only for some domains and for some $\lambda(t)$.

**Problem 2.** Let $D$ be the half plane $y > 0$ and $\lambda(t) = 1$. Then one has the Dirichlet problem

$$\operatorname{Re}[W(t)] = g(t), \quad t \in R,$$

so that

$$W^+(t) + \overline{W^+(t)} = 2g(t), \quad t \in R. \tag{12}$$

Multiply (12) by $\dfrac{1}{2\pi i(t-z)}$, for $z \in D$, and then integrate on $R$. By the Cauchy integral formula one has

$$W(z) = \frac{1}{\pi i} \int_R \frac{g(t)dt}{t-z} + ic. \tag{13}$$

The real part of (13) defines a harmonic function in $D$ that vanishes at infinity as the solution of the Dirichlet problem

$$u(x,y) = \frac{y}{\pi} \int_R \frac{g(t)dt}{(t-x)^2 + y^2}. \tag{14}$$

Now let $D$ be the circular domain $|z| < 1$, where $|z| = 1$ is denoted by $\Gamma$. Then from (12) in the same way as above one can obtain

$$W(z) + \overline{W(0)} = \frac{1}{\pi i} \int_\Gamma \frac{g(t)dt}{t-z}, \tag{15}$$

and, consequently,

$$W(0) + \overline{W(0)} = \frac{1}{\pi i} \int_\Gamma \frac{g(t)dt}{t}.$$

Thus (15) can be written as

$$W(z) = \frac{1}{2\pi i} \int_\Gamma \frac{g(t)(t+z)}{t-z} \frac{dt}{t} + ic, \quad |z| < 1, \tag{16}$$

which is Schwartz's well-known formula. The solution of Dirichlet's problem for a harmonic function in $D$ will be the real part of $W(z)$:

I. Two-Dimensional Cases

$$u(x,y) = \frac{1-\rho^2}{2\pi} \int_0^{2\pi} \frac{g(t)d\varphi}{1-2\rho\cos(\varphi-\psi)+\rho^2}, \quad z = x+iy = \rho e^{i\varphi}, \quad (17)$$

which is Poisson's well-known formula.

Using Schwartz's formula we can construct the solution of the Neumann problem: find a harmonic function in $D: |z| < 1$ that satisfies the conditions

$$\frac{\partial u}{\partial \rho} = f(t), \quad \rho = |t| = 1. \quad (17_1)$$

**Solution.** If $u$ is harmonic in $D$, then $r\dfrac{\partial u}{\partial r}$ is harmonic too and satisfies $(17_1)$ on the boundary. It is easy to see that if the direction of the normal is inside of $\Gamma$, then

$$r\frac{\partial u}{\partial r} = -\frac{\partial u}{\partial x}x - \frac{\partial u}{\partial y}y = -\mathrm{Re}\left[\left(\frac{\partial u}{\partial x} - i\frac{\partial u}{\partial y}\right)(x+iy)\right] = -\mathrm{Re}[\phi'(z)z].$$

Thus for the holomorphic function $z\phi'(z)$ one has the Dirichlet problem $(17_1)$, which can be represented as in (16):

$$-z\phi'(z) = \frac{1}{2\pi i} \int_\Gamma \frac{f(t)(t+z)}{t-z}\frac{dt}{t}.$$

Because the solution of the Neumann problem must satisfy

$$\int_\Gamma f(t)ds = \int_0^{2\pi} f(t)d\vartheta = 0,$$

we can write

$$z\phi'(z) = -\frac{1}{2\pi} \int_0^{2\pi} f(t)\frac{2z}{t-z}d\vartheta,$$

$$\phi(z) = \frac{1}{2\pi} \int_0^{2\pi} f(t)\ln(t-z)^2 d\vartheta.$$

Correspondingly its real part will be the solution of $(17_1)$:

$$u(x, y) = \frac{1}{2\pi} \int_0^{2\pi} f(t) \ln(1 + r^2 - 2r\cos(\vartheta - \psi)) d\vartheta, \quad z = re^{-\psi i}.$$

This is called the Dini integral.

**Problem 3.** The solution of the problem (8), (9) can be obtained in a simple case too. Let $D^+$ be the circular domain $|z| < 1$, $\Gamma_0 : |z| = 1$, $\lambda(t) = 1$. Then the solution of (9) can be represented as (it is possible to consider for general $G(t)$)

$$W(z) = \frac{\chi(z)}{2\pi i} \int_\Gamma \frac{g_2(t) dt}{\chi^+(t)(t-z)} + \phi(z)\chi(z), \quad |z| < 1,$$

where $\phi(z)$ is a holomorphic function inside $\Gamma_0$, $\chi(z)$ a canonical function for (9), and $\Gamma = U_1^n \Gamma_k$. Putting this representation in (8) for $\phi(z)$ one will obtain the Dirichlet problem. Thus it will be represented by the Schwartz formula (16).

If $\Gamma_0$ is the real axis and $D^+$ the half plane, the problem (8), (9) can be solved too. The Riemann–Schwartz principle of reflection is used frequently below.

Let $D$ be a domain in the plane $z = x + iy$ the boundary of which contains the segment $L$ of real axis $y = 0$ or the arc $\Gamma$ of the circle $|z| = 1$. Let $\phi(z) = u + iv$ be a holomorphic function in $D$ and

$$u(x, 0) = 0 \quad \text{or} \quad v(x, 0) = 0, \quad x \in L. \tag{$18_1$}$$

Then the function

$$\psi(z) = \begin{cases} \phi(z), & z \in D, \\ \mp \overline{\phi(\bar{z})}, & z \in D^*, \end{cases}$$

where $D^*$ is a symmetric domain of $D$ with respect to $L$, is holomorphic in $D \cup D^* \cup L$. The sign $(-)$ is for the first condition in $(18_1)$ and $(+)$ for the second one. But if we have the conditions

$$u(x, y) = 0 \quad \text{or} \quad v(x, y) = 0, \quad x + iy = t \in \Gamma,$$

then

$$\psi(z) = \begin{cases} \phi(z), & z \in D, \\ \mp\overline{\phi(z^*)}, & \\ z \in D^*, & z^* = \dfrac{1}{\bar{z}}, \end{cases}$$

will be holomorphic in $D \cup D^* \cup \Gamma$. In this case $D^*$ is a symmetric domain of $D$ with respect to $\Gamma$.

**Problem 4.** Let the domain $|z| < 1$, $y > 0$ be denoted by $D$ and have boundary conditions

$$\text{Re}[W(t)] = f(t), \quad |z| = 1, \ y > 0, \tag{18}$$
$$\text{Re}[W(x)] = \varphi(x), \quad -1 < x < 1. \tag{19}$$

One can consider $\varphi(x) = 0$ without loss of generality because, by using the solution of the Dirichlet problem for the half plane $y > 0$, condition (19) can be reduced to the homogeneous case. Then the function

$$\phi(z) = \begin{cases} W(z), & z \in D, \\ -\overline{W(\bar{z})}, & \\ y < 0, & |z| < 1, \end{cases}$$

by force of the Riemann–Schwartz principle of reflection, is holomorphic in $|z| < 1$ and, by (18), satisfies the condition

$$\text{Re}[\phi(t)] = \begin{cases} f(t), & \\ |t| = 1, \ y > 0, \\ -\overline{f(\bar{t})}, & \\ |t| = 1, \ y < 0. \end{cases}$$

Thus it will be represented as in (16). If in place of (19) one has

$$\text{Im}[W(t)] = 0, \quad -1 < t < 1,$$

then the function

$$\phi(z) = \begin{cases} W(z), \\ |z| < 1, \quad y > 0, \\ \overline{W(\bar z)}, \\ |z| < 1, \quad y < 0, \end{cases}$$

is holomorphic in $|z| < 1$ and the boundary condition for $|t| = 1$, $y < 0$ correspondingly is defined. Thus the solution will be represented with the help (16).

**Problem 5.** Keldish–Sedov mixed problem. Let $D$ be the half plane, $L_1 : [a_1, b_1]$, ..., $[a_n, b_n]$, $(a_1 < b_1 < a_2 < \cdots < a_n < b_n)$ a set of segments on the real axis, and $L_2$ the remaining part of real axis.

Define a holomorphic function $W = u + iv$ in $D$ that vanishes at infinity by the conditions

$$u^+(x) = f(x), \quad x \in L_1, \quad v^+(x) = \varphi(x), \quad x \in L_2. \tag{20}$$

Suppose that $W(z)$ can have integrable singularities at the ends of these segments

$$|W(z)| < \frac{const}{|z-c|^\alpha}, \quad \alpha = const < 1. \tag{$20_1$}$$

This condition will always be supposed for domains with cuts along some segments.

To solve the BVP for the domain $D$ with cuts along the open smooth lines $L_k : [a_k, b_k]$, $k = 1, \ldots, n$, $L = \cup_1^n L_k$, consider the function

$$\chi(z) = \prod_1^n (z - a_k)^{-\gamma} (z - b_k)^{\gamma - 1},$$

where $\gamma = \alpha + i\beta$ is constant. The branch of this function is fixed in such a way that

$$\lim_{|z| \to \infty} z^n \chi(z) = 1.$$

From left side of $L$, $\chi^+(t) = \chi(t)$, $t \in L$. Then from the right side of $L$

$$\chi^-(t) = e^{-2\pi i \gamma} \chi(t).$$

Thus

# I. Two-Dimensional Cases

$$\chi^+(t) = a\chi^-(t), \quad t \in L, \quad a = e^{2\pi i \gamma}, \tag{20_2}$$

$$\gamma = \frac{1}{2\pi i} \ln |a| + \frac{1}{2\pi}\theta, \quad \theta = \arg a,$$

where

$$0 \le \theta < 2\pi, \quad \text{i.e.,} \quad 0 \le \alpha < 1,$$

and $(z-a_k)(z-b_k)\chi(z)$ will be bounded on the points $a_k, b_k$. Thus $\chi(z)$ is holomorphic function in $D$.

In the following we assume: (a) the boundary conditions on $L$ are satisfied on all points of $L$ except the endpoints $a_k, b_k, k = 1, \ldots, n$, and (b) the holomorphic function $\varphi(z)$ in $D$ is continuously continued on $L$ everywhere except possibly on the endpoints $a_k, b_k$, where $\varphi(z)$ has the property $(20_1)$.

For the piecewise holomorphic function

$$\phi(z) = \begin{cases} W(z), & y > 0, \\ \overline{W(\bar{z})}, & y < 0, \end{cases} \quad (a)$$

the conditions (20) can be written as

$$\phi^+ + \phi^- = 2f(x), \quad x \in L_1, \tag{21}$$
$$\phi^+ - \phi^- = 2i\varphi(x), \quad x \in L_2. \tag{22}$$

The solution of this problem can be represented as

$$\phi(z) = \frac{1}{\pi} \int_{L_2} \frac{\varphi(t)dt}{t-z} + \frac{R(z)}{\pi i} \int_{L_1} \frac{f(t)dt}{R^+(t)(t-z)}, \tag{23}$$

$$R_k(z) = \sqrt{\frac{z-a_k}{z-b_k}}, \quad R(z) = \prod_1^n R_k(z), \quad R^+(x) = R(x).$$

One can guarantee that $R^+(x) = -R^-(x)$, $x \in L_1$, $R^+(x) = R^-(x)$, $x \in L_2$. This solution is bounded on $a_k$ and unbounded on $b_k$. Because $\overline{R(\bar{z})} = R(z)$ and $R^+ = \overline{R^-(x)} = -\overline{R^+}$ on $L_1$, the solution (23) satisfies the condition

$$\overline{\phi(\bar{z})} = \phi(z), \quad (b).$$

**Problem 6.** Let $D$ be the $z$ plane with the cuts $(a_k, b_k)$, $k = 1, \ldots, n$, along the real segments. The set of these segments will be denoted by $L$.

Find the holomorphic function $W(z) = u + iv$ in $D$ that vanishes at infinity by the conditions

$$u^+(x) = f^+(x), \quad u^-(x) = f^-(x), \quad x \in L. \tag{24}$$

Define the holomorphic functions

$$\begin{aligned} \phi_1(z) &= \frac{1}{2}[W(z) + \overline{W(\bar{z})}], \\ \phi_2(z) &= \frac{1}{2}[W(z) - \overline{W(\bar{z})}]. \end{aligned} \tag{25}$$

Then the conditions (24) can be written as

$$\begin{aligned} \phi_1^+ + \phi_1^- &= [f^+ + f^-] \equiv 2f, \quad x \in L, \\ \phi_2^+ - \phi_2^- &= [f^+ - f^-] \equiv 2\varphi. \end{aligned}$$

Thus by force of (23), $\phi_1$ and $\phi_2$ are

$$\begin{aligned} \phi_1(z) &= \frac{R(z)}{\pi i} \int_L \frac{f(t)dt}{R(t)(t-z)}, \\ \phi_2(z) &= \frac{1}{\pi i} \int_L \frac{\varphi(t)dt}{t-z}, \end{aligned} \tag{26}$$

and W(z) is defined by (25), (26).

From (25) one has conditions

$$\overline{\phi_1(\bar{z})} = \phi_1(z), \quad \overline{\phi_2(\bar{z})} = -\phi_2(z)$$

which are satisfied by force of (b).

**Problem 7.** Let $D$ be the $z$ plane with cuts along segments of the circle $|z| = 1$, $L : [a_k, b_k]$, $k = 1, \ldots, n$. Find the holomorphic function $W(z) = u + iv$ in $D$ that vanishes at infinity by the conditions

$$u^+(t) = f^+(t), \quad u^-(t) = f^-(t), \quad t \in L. \tag{27}$$

14    I. Two-Dimensional Cases

Consider the two holomorphic functions in $D$

$$\phi_1(z) = \frac{1}{2}\left[W(z) + \overline{W\left(\frac{1}{\bar{z}}\right)}\right],$$

$$\phi_2(z) = \frac{1}{2}\left[W(z) - \overline{W\left(\frac{1}{\bar{z}}\right)}\right].$$

Conditions (27) give

$$\phi_1^+ + \phi_1^- = f^+ + f^- \equiv 2f(t),$$
$$\phi_2^+ - \phi_2^- = f^+ - f^- \equiv 2\varphi(t), \quad x \in L,$$

which can be represented as

$$\phi_1(z) = \frac{R(z)}{\pi i}\int_L \frac{f(t)dt}{R^+(t)(t-z)},$$

$$\phi_2(z) = \frac{1}{\pi i}\int_L \frac{\varphi(t)dt}{t-z},$$

$$R_k(z) = \sqrt{\frac{z - a_k}{z - b_k}},$$

where $W(z) = \phi_1(z) + \phi_2(z)$ and $R(z) = \prod_1^n R_k(z)$. It is easy to see that

$$\overline{\phi_1\left(\frac{1}{\bar{z}}\right)} = \phi_1(z), \quad \overline{\phi_2\left(\frac{1}{\bar{z}}\right)} = -\phi_2(z).$$

The above considered problems will be formulated for the generalized Cauchy–Riemann system

$$\frac{\partial W}{\partial \bar{z}} + B(z)\overline{W} = 0, \tag{28}$$

for Beltrami and generalized Beltrami equations

$$\frac{\partial W}{\partial \bar{z}} + q\frac{\partial W}{\partial z} = 0, \ |q| < 1, \quad \frac{\partial W}{\partial \bar{z}} + q_1\frac{\partial W}{\partial z} + q_2\frac{\partial \overline{W}}{\partial \bar{z}} = 0, \ |q_1| + |q_2| < 1, \tag{29}$$

and the solutions will be represented in quadrature only in some partial cases. This chapter considers correctly-posed BVP for the polyregular equation

$$\frac{\partial^n W}{\partial \bar{z}^n} = 0, \quad n \geq 2, \tag{30}$$

and in partial cases, their solutions are represented in quadratures.

If $W$ is a solution of (30), then it is a solution of the polyharmonic equation

$$\Delta^n W = 0. \tag{31}$$

The general solution of (30) can be written as

$$W = \sum_1^n \bar{z}^{k-1} \varphi_k(z), \tag{32}$$

where the $\varphi_k(z)$ are holomorphic functions. Note that for $n = 2$, (30) is called Bitsadze's equation for which the homogeneous problem on the circle

$$W(t) = 0, \quad |t| = 1, \tag{33}$$

has, as was showed by Bitsadze, infinitely many solutions

$$W(z) = (1 - z\bar{z})\varphi(z),$$

where $\varphi(z)$ is any holomorphic function. It is clear that the nonhomogeneous problem $W(t) = f(t)$, $|t| = 1$ is ill posed. In general, for holomorphic functions, when the real and imaginary parts are given on the boundary, the problem is not correctly posed. For (30) correctly-posed BVP are those that are correctly posed for polyharmonic functions. For example, the classical Dirichlet problem and the Riquie problem are correctly posed.

For plane isotropic elasticity some BVP will be solved using Kolosov–Muskhelishvili representations for the displacements $u, v$ and for the stresses [Mu2]

$$\begin{aligned} 2\mu(u + iv) &= \varkappa\varphi(z) - \overline{z\varphi'(z)} - \overline{\psi(z)}, \\ i(X + iY) &= \varphi(z) + \overline{z\varphi'(z)} + \overline{\psi(z)}, \end{aligned} \tag{34}$$

where $X + iY = \int_{AB} (X_n + iY_n) ds$ is the resultant vector of the forces acting on an arc $AB$, where $A$ is a fixed point, $B$ is a variable point, $\lambda, \mu$ are elasticity constants, $\varphi(z), \psi(z)$ are holomorphic functions, and $\varkappa$ is Muskhelishvili's constant.

# I. Two-Dimensional Cases

**Problem 8.** Let $D$ be the circular domain $|z| < 1$ or half plane $y > 0$. Find the equilibrium of $D$ by the condition

$$2\mu(u+iv) = g(t) = g_1 + ig_2.$$

**Solution.** Let $\Gamma$ be $|z| = 1$. Then by force of (34) we have

$$\begin{aligned}\operatorname{Re}[\varkappa\varphi(t) - \bar{t}\varphi'(t) - \psi(t)] &= g_1(t),\\ \operatorname{Im}[\varkappa\varphi(t) + \bar{t}\varphi'(t) + \psi(t)] &= g_2(t), \quad t \in \Gamma.\end{aligned} \quad (35)$$

Consider the holomorphic functions in $D$

$$\begin{aligned}\phi_1(z) &= \varkappa\varphi(z) - \frac{1}{z}\varphi'(z) - \psi(z),\\ \phi_2(z) &= \varkappa\varphi(z) + \frac{1}{z}\varphi'(z) + \psi(z),\end{aligned} \quad (36)$$

where $\varphi'(0) = 0$. Then by force of (35) one has

$$\operatorname{Re}\phi_1(t) = g_1(t), \quad \operatorname{Im}[\phi_2(t)] = -\operatorname{Re}[i\phi_2(t)] = g_2(t), \quad t \in \Gamma. \quad (37)$$

Thus $\phi_1(z)$, $\phi_2(z)$ are defined by (16). Then from (36) $\varphi(z)$ and $\psi(z)$ are easily defined

$$\begin{aligned}\varphi(z) &= \frac{1}{2\varkappa}[\phi_1(z) + \phi_2(z)],\\ \psi(z) &= \frac{1}{2}[\phi_2(z) - \phi_1(z)] - \frac{1}{z}\varphi'(z).\end{aligned} \quad (38)$$

If $D$ is the half plane $y > 0$, then by force of (35) the holomorphic functions in $D$

$$\begin{aligned}\phi_1(z) &= \varkappa\varphi(z) - z\varphi'(z) - \psi(z),\\ \phi_2(z) &= \varkappa\varphi(z) + z\varphi'(z) + \psi(z),\end{aligned} \quad (39)$$

satisfy conditions (37) and $\phi_1$, $\phi_2$ are defined by (13) so that $\varphi(z)$, $\psi(z)$ are known. If the stresses are given on $\Gamma$ the problem can be solved in an analogous way.

**Problem 9.** Let $D$ be the half plane $y > 0$. $L_1$ is the set of segments $[a_k, b_k]$, $k = 1, \ldots, n$, and $L_2$ is the remaining part of the $x$ axis, where $L = L_1 \cup L_2$ is the $x$ axis. Find the equilibrium of an elastic body $D$ by the conditions

$$\begin{aligned}2\mu(u+iv)^+ &= f, \quad x \in L_1,\\ (X_n + iY_n)^+ &= g, \quad x \in L_2.\end{aligned} \quad (40)$$

It is supposed that $u$, $v$ vanish at infinity.

**Solution.** Consider the piecewise holomorphic function

$$\phi(z) = \begin{cases} \varphi(z), & y > 0, \\ -\overline{z\varphi'(\bar z)} - \overline{\psi(\bar z)}, & y < 0. \end{cases} \tag{41}$$

Then by force of (34), the boundary conditions (40)

$$\varkappa\phi^+ + \phi^- = f(x), \quad x \in L_1,$$
$$\phi^+ - \phi^- = g(x), \quad x \in L_2.$$

Thus the solution has the representation

$$\phi(z) = \frac{1}{2\pi i}\int_{L_2}\frac{g(t)dt}{t-z} + \frac{\chi(z)}{2\pi i}\int_{L_1}\frac{f(t)dt}{\chi^+(t)(t-z)},$$

where the canonical function $\chi(z)$ is defined by $(20_2)$.

**Problem 10.** Find the equilibrium of $D$ by the conditions

$$2\mu(u+iv)^+ = f_1 + if_2, \quad x \in L_1, \tag{42}$$
$$v^+ = f(x), \quad X_n^+ = g(x), \quad x \in L_2. \tag{43}$$

In other words, on $L_2$ one has stamps without friction.

**Solution.** Because $v(x)$ is given on all $L$, by force of (36) one has

$$\operatorname{Im}[\varkappa\varphi(x) + x\varphi'(x) + \psi(x)] = f(x), \quad x \in L.$$

By this condition the holomorphic function in $D$

$$\phi(z) = \varkappa\varphi(z) + z\varphi'(z) + \psi(z) \tag{44}$$

can be defined as the solution of the Dirichlet problem (13)

$$\operatorname{Re}[i\phi(x)] = -f(x), \quad x \in L.$$

Defining $z\varphi'(z) + \psi(z)$ from (44) and substituting into the equalities
$$u^+ = \operatorname{Re}[\varkappa\varphi(t) - t\varphi'(t) - \psi(t)], \quad x \in L_1,$$
$$X^+ = \operatorname{Re}[\varphi(t) + t\varphi'(t) + \psi(t)], \quad x \in L_2,$$

gives

$$\operatorname{Re}[\varphi(t)] = F(t), \quad t \in L. \tag{45}$$

Thus $\varphi(z)$ is represented by (13) once again.

If one has stamps with friction on $L_2$, then the second condition of (43) must be changed by

$$Y_n = kX_n, \quad \text{i.e.,} \quad Y = kX,$$

and for $\varphi(z)$, as above, we have the Dirichlet problem (45). $\psi(z)$ is correspondingly defined by (44).

# 1   BVP for Holomorphic Functions

Let $D$ be the plane $z = x + iy$ with cuts along the segments $L_1, L_2$ ($L = L_1 \cup L_2$) of the real axis. Let $D^+$ be the half plane $x > 0$ with cuts along the segments $L$ of the $x$ axis. And let $D_1$ be the $z$ plane with cuts along the segments $[-a, a]$ of the $y$ axis and $[a_k, b_k]$, $k = 1, \ldots, n$, of the $x$ axis.

**Problem 1.** Find the holomorphic function in $D$ $w(z) = u + iv$ that vanishes at infinity by the conditions

$$\begin{aligned} u^+ &= f_1^+(x), & u^- &= f_1^-(x), & x \in L_1, \\ v^+ &= f_2^+(x), & v^- &= f_2^-(x), & x \in L_2. \end{aligned} \tag{1.1}$$

**Solution.** Consider the two holomorphic functions

$$\phi_1(z) = \frac{1}{2}[w(z) + \overline{w(\bar{z})}], \ \phi_2(z) = \frac{1}{2}[w(z) - \overline{w(\bar{z})}]. \tag{1.2}$$

Then (1.1) can be written as

$$\begin{aligned} \phi_1^+ + \phi_1^- &= f_1^+ + f_1^- \equiv 2f_1(x), & x \in L_1, \\ \phi_1^+ - \phi_1^- &= i(f_2^+ - f_2^-) \equiv 2if_2(x), & x \in L_2, \\ \phi_2^+ + \phi_2^- &= i(f_2^+ + f_2^-) \equiv 2i\varphi_1(x), & x \in L_2, \\ \phi_2^+ - \phi_2^- &= f_1^+ - f_1^- \equiv 2\varphi_2(x), & x \in L_1. \end{aligned}$$

By force of (23) these functions are represented

$$\phi_1(z) = \frac{R(z)}{\pi i} \int_{L_1} \frac{f_1(t)dt}{R(t)(t-z)} + \frac{1}{\pi} \int_{L_2} \frac{f_2(t)dt}{t-z},$$
$$\phi_2(z) = \frac{R(z)}{\pi} \int_{L_2} \frac{\varphi_1(t)dt}{R(t)(t-z)} + \frac{1}{\pi i} \int_{L_1} \frac{\varphi_2(t)dt}{t-z},$$
(1.3)

and $w(z)$ will be defined by (1.2). It is easy to see that $\overline{\phi_1(\bar{z})} = \phi_1(z), \overline{\phi_2(\bar{z})} = -\phi_2(z)$ since $\overline{R^+(t)} = R^+(t), t \in L_2$ and $\overline{R^+(t)} = -R(t), t \in L_1$.

**Problem 2.** Find the holomorphic function in $D$ $w(z) = u + iv$ that vanishes at infinity by the conditions

$$u^+ = f^+(x), \quad v^- = f^-(x), \quad x \in L = L_1 \cup L_2. \tag{1.4}$$

**Solution.** Consider the two holomorphic functions in $D$

$$\Phi_1(z) = w(z) - i\overline{w(\bar{z})},$$
$$\Phi_2(z) = w(z) + i\overline{w(\bar{z})},$$
(1.5)

that satisfy

$$\overline{\Phi_1(\bar{z})} = i\Phi_1(z), \quad \overline{\Phi_2(\bar{z})} = -i\Phi_2(z). \tag{1.6}$$

By force of (1.4) one has

$$\Phi_1^+ + i\Phi_1^- = 2[f^+ - f^-] \equiv f_1(x),$$
$$\Phi_2^+ - i\Phi_2^- = 2[f^+ + f^-] \equiv 2f_2(x), \quad x \in L.$$

By force of ($20_2$) the solution of this problem is

$$\Phi_k(z) = \frac{X_k(z)}{2\pi i} \int_L \frac{f_k(t)\,dt}{X_k^+(t)(t-z)}, \quad k = 1, 2, \tag{1.7}$$

where

$$X_k(z) = \prod_1^n (z-a_j)^{-\gamma_k}(z-b_j)^{\gamma_k-1}, \quad \gamma_1 = \frac{1}{4}, \quad \gamma_2 = \frac{3}{4}.$$

20   I. Two-Dimensional Cases

It is easy to verify that $X_k(z) = \overline{X_k(\bar{z})}$. In any case if $R(z) = \sqrt[n]{z}$, then $R(\bar{z}) = \sqrt[n]{|z|}\, e^{-\frac{i\omega}{n}}$, i.e., $\overline{R(\bar{z})} = R(z)$. Thus $\overline{X_1^+(t)} = X_1^- = iX_1^+$ and $\overline{X_2^+(t)} = X_2^- = -iX_2^+$. By force of these equalities (1.7) satisfies (1.6). The unknown function $w(z)$ is defined by (1.6).

**Problem 3.** Find the holomorphic function $w(z)$ in $D^+$ that vanishes at infinity by the conditions

$$u(0, y) = f(y), \quad y \in R.$$

Conditions (1.1) or (1.4) hold on the segments $L$ of the $x$ axis.

Without loss of generality the condition on the $y$ axis can be considered homogeneous: $u(0, y) = 0$. Then by the Riemann–Schwartz principle of reflection the function

$$\Phi(z) = \begin{cases} w(z), & x > 0, \\ -\overline{w(-\bar{z})}, & x < 0, \end{cases}$$

is holomorphic in the $z$ plane with cuts along the segments $L$, $L^*$ of the real axis, where $L^*$ is the reflection of $L$ with respect to the $y$ axis. On $L^*$ the boundary conditions are defined by the conditions on $L$. Thus we obtain $\Phi(z)$ in the above considered Problem 1, which was solved in quadratures.

**Problem 4.** Let $D$ be the $z$ plane with cuts $L : [a_k, b_k]$, $k = 1, \ldots, n$, along the arc segments of the circle $|z| = 1$. Find the holomorphic function in $D$ that vanishes at infinity by the conditions

$$u^+(t) = f^+(t), \quad v^-(t) = f^-(t), \quad t \in L, \tag{1.8}$$

or

$$\left. \begin{array}{l} u^+(t) = f^+(t), \quad u^-(t) = f^-(t), \quad t \in L_1, \\ v^+(t) = \varphi^+(t), \quad v^-(t) = \varphi^-(t), \quad t \in L_2. \end{array} \right\} \quad L = L_1 \cup L_2. \tag{1.9}$$

**Solution.** Consider the two holomorphic functions in the case of (1.9)

$$\Phi_1(z) = \frac{1}{2}\left[w(z) + \overline{w\left(\frac{1}{\bar{z}}\right)}\right], \quad \Phi_2(z) = \frac{1}{2}\left[w(z) - \overline{w\left(\frac{1}{\bar{z}}\right)}\right] \tag{1.10}$$

and in the case of (1.8)

$$\Phi_1(z) = w(z) - i\overline{w\left(\frac{1}{\bar{z}}\right)}, \quad \Phi_2(z) = w(z) + i\overline{w\left(\frac{1}{\bar{z}}\right)}, \tag{1.11}$$

Then (1.8), (1.9) provide conditions analogous to those for (1.1) and (1.4). Thus they are defined too.

**Problem 5.** Let $D$ be the circle $|z| < 1$ with cuts along the segments $L$ of the $x$ axis. Let $D^+$ be the half circle $|z| < 1$, $x > 0$ with cuts along the segments $L$ of the $x$ axis.
(a) Find the holomorphic function in $D$ by conditions

$$u^+(t) = 0, \quad |t| = 1.$$

Conditions on $L$ are as for Problems 1 or 2.
By the Riemann–Schwartz principle of reflection

$$\Phi(z) = \begin{cases} w(z), & |z| < 1, \\ -\overline{w\left(\frac{1}{\bar{z}}\right)}, & |z| > 1, \end{cases} \tag{1.12}$$

is a holomorphic function in the $z$ plane bounded at $\infty$ with cuts along the segments $L$, $L^*$ of the $x$ axis, where $L^*$ is the reflection of $L$ with respect to the circle $|z| = 1$. On $L^*$ the boundary conditions are defined by the conditions on $L$. Thus we obtain Problem 1 or 2.
(b) Find the holomorphic function in $D^+$ by the condition

$$u^+(0, y) = 0. \tag{1.13}$$

Conditions on the segments and on $|z| = 1$, $x > 0$ are as for Problems 1, 2, or 3, 5.

**Solution.** By the Riemann–Schwartz principle of reflection and by force of (1.13) the function

$$\Phi(z) = \begin{cases} w(z), & x > 0, \ |z| < 1, \\ -\overline{w(-\bar{z})}, & x < 0, \ |z| < 1, \end{cases} \tag{1.14}$$

is holomorphic for $|z| < 1$ with cuts along the segments $L$, $L^*$ of the $x$ axis. Thus this case is reduced to the case of (a).

22    I. Two-Dimensional Cases

**Problem 6.** Let $D$ be the half plane $y > 0$ with cuts along the arc segments $L$ of the half circle $|z| = 1$, $y > 0$. Find the holomorphic function in $D$ that vanishes at infinity by the condition

$$u^+(x, 0) = 0, \quad x \in R.$$

Conditions on $L$ are as for Problem 5. Again by the Riemann–Schwartz principle of reflection the function

$$\Phi(z) = \begin{cases} w(z), & y > 0, \\ -\overline{w(\bar{z})}, & y < 0, \end{cases} \quad (1.14_1)$$

is as in Problem 5, where the arc segments of the circle consist of $L$ and its reflection $L^*$. Conditions on $L^*$ are defined by the conditions on $L$.

**Problem 7.** Let $D$ be the domain with the boundary $y = 0$, $|x| > 1$ and $|z| = 1$, $y \geq 0$, and with cuts of the $y$ axis for $y > 1$. Find the holomorphic function in $D$ that vanishes at infinity by the conditions

$$u(t) = 0, \quad |t| = 1, \quad y > 0. \quad (1.15)$$

Conditions on the boundary $y = 0$, $|x| > 1$ and on the cuts are as for Problem 3.

**Solution.** Consider the function

$$\Phi(z) = \begin{cases} w(z), & |x| > 1, \ y > 0, \\ -\overline{w\left(\frac{1}{\bar{z}}\right)}, & |x| < 1, \ y > 0. \end{cases}$$

By force of (1.15) it will be a holomorphic function in the half plane $y > 0$ with cuts along the $y$ axis. By the conditions on $y = 0$, $x > 1$ and on the cuts, boundary conditions are defined. Thus this problem is solved like Problem 3.

The Riemann–Schwartz principle of reflection can also be successfully used to define holomorphic functions in the quarter plane ($x > 0$, $y > 0$) with corresponding boundary conditions. For instance,

$$\operatorname{Re} \Phi(x, 0) = f(x), \quad x > 0,$$
$$\operatorname{Re} \Phi(0, y) = 0 \quad \text{or} \quad \operatorname{Im} \Phi(0, y) = 0, \quad y > 0.$$

## 1. BVP for Holomorphic Functions

If $w(z)$ is a solution of the nonhomogeneous equation

$$\frac{\partial w}{\partial \bar{z}} = F(z),$$

then all the above problems for holomorphic functions can be solved explicitly using the partial solution of this equation defined by ($5_1$)

$$w(z) = -\frac{1}{\pi} \iint_D \frac{F(\zeta)\,d\zeta}{\zeta - z}.$$

If $D$ is the $z$ plane with cuts $L$ of the $x$ axis, then the BVP of plane isotropic elasticity can be solved.

**Problem 8.** Find the equilibrium of $D$ with the condition

$$2\mu(u^\pm + iv^\pm) = f^\pm(x), \quad x \in L, \tag{1.16}$$

or

$$X_n^\pm + iY_n^\pm = g^\pm(x). \tag{1.17}$$

Assume that the displacement vanishes at infinity.

**Solution.** By force of (1.16) and (34) one has

$$\operatorname{Re}\left[\varkappa\varphi(x) - x\varphi'(x) - \psi(x)\right]^\pm = 2\mu u^\pm,$$
$$\operatorname{Im}\left[\varkappa\varphi(x) + x\varphi'(x) + \psi(x)\right]^\pm = 2\mu v^\pm.$$

Consider the two holomorphic functions in $D$

$$\begin{aligned}\Phi_1(z) &= \varkappa\varphi(z) - z\varphi'(z) - \psi(z),\\ \Phi_2(z) &= \varkappa\varphi(z) + z\varphi'(z) + \psi(z).\end{aligned} \tag{1.18}$$

Conditions (1.16) are written as

$$\begin{aligned}\operatorname{Re} \Phi_1^\pm(x) &= f_1^\pm,\\ \operatorname{Im} \Phi_2^\pm(x) &= -\operatorname{Re}\left[i\Phi_2^\pm(x)\right] = f_2^\pm(x), \quad x \in L.\end{aligned} \tag{1.19}$$

Thus $\Phi_1$ and $\Phi_2$ are defined as for Problem 1.

## I. Two-Dimensional Cases

In the case of (1.17) by force of (34) we have

$$\operatorname{Re}\left[\varphi(x) + x\varphi'(x) + \psi(x)\right]^{\pm} = f_1^{\pm},$$
$$\operatorname{Im}\left[\varphi(x) - x\varphi'(x) - \psi(x)\right]^{\pm} = f_2^{\pm}, \quad x \in L.$$

For the two holomorphic functions

$$\begin{aligned}\Phi_1(z) &= \varphi(z) + z\varphi'(z) + \psi(z),\\ \Phi_2(z) &= \varphi(z) - z\varphi'(z) - \psi(z),\end{aligned} \quad (1.20)$$

(1.17) gives again boundary conditions like (1.19). Knowing $\Phi_1$, $\Phi_2$ gives $\varphi(z)$, $\psi(z)$ via (1.18) and (1.20).

**Problem 9.** Find the equilibrium of $D$ with the conditions

$$\begin{aligned}2\mu(u^+ + iv^+) &= f^+(x),\\ X_n^- + iY_n^- &= g^-(x), \quad x \in L.\end{aligned} \quad (1.21)$$

**Solution.** This problem was considered by Sherman and solved in an interesting way by Muskhelishvili (see [Mu2]).

Consider the holomorphic functions

$$\Phi_1(z) = \varphi(z), \quad \Phi_2(z) = \overline{z\,\varphi'(\bar z) + \psi(\bar z)}. \quad (1.22)$$

Conditions (1.21) are written as

$$\begin{aligned}\varkappa\Phi_1^+ - \Phi_2^- &= f^+(x),\\ \Phi_1^- + \Phi_2^+ &= f^-(x), \quad x \in L.\end{aligned}$$

From these equalities one can obtain

$$\left[\sqrt{\varkappa}\Phi_1 + i\Phi_2\right]^+ + \frac{i}{\sqrt{\varkappa}}\left[\sqrt{\varkappa}\Phi_1 + i\Phi_2\right]^- = \frac{1}{\sqrt{\varkappa}}f^+ + if^- \equiv f(x),$$

$$\left[\sqrt{\varkappa}\Phi_1 - i\Phi_2\right]^+ - \frac{i}{\sqrt{\varkappa}}\left[\sqrt{\varkappa}\Phi_1 - i\Phi_2\right]^- = \frac{1}{\sqrt{\varkappa}}f^+ - if^- \equiv g(x), \quad x \in L.$$

With these boundary conditions and $(20_2)$ we can define the functions $\sqrt{\varkappa}\Phi_1(z) + i\Phi_2(z)$ and $\sqrt{\varkappa}\Phi_1(z) - i\Phi_2(z)$ and correspondingly the functions $\varphi(z)$, $\psi(z)$ are defined by (1.22).

In the case where $D$ is the $z$ plane with cuts along the arcs of the circle $|z| = 1$, the above problems can be solved using the piecewise functions

$$\Phi_1(z) = \varphi(z), \quad \Phi_2(z) = \frac{1}{z}\overline{\varphi\left(\frac{1}{\bar z}\right)} + \overline{\psi\left(\frac{1}{\bar z}\right)}.$$

1. BVP for Holomorphic Functions    25

**Problem 10.** Find the equilibrium of $D$ if one of the following conditions is given:

1) $2\mu(u^+ + iv^+) = f^+$,  $v^- = g^-$,  $X_n^- = \varphi$,  $x \in L$, (1.23)
2) $2\mu(u + iv)^\pm = f^\pm$,  $x \in L_1$,
   $v^\pm = g^\pm$,  $X_n^\pm = \varphi^\pm$,  $x \in L_2$,  $L_1 \cup L_2 = L$. (1.24)

**Solution.** Because $v^\pm$ is given on all $L$, by force of (34) we have

$$\text{Im}\left[\varkappa\varphi(x) + x\varphi'(x) + \psi(x)\right]^\pm = g^\pm, \quad x \in L.$$

Thus the holomorphic function in $D$

$$\Phi(z) = \varkappa\varphi(z) + z\varphi'(z) + \psi(z) \tag{1.26}$$

is defined as for Problem 1. Then from (1.26) define the function $z\varphi'(z) + \psi(z)$ and put it into (1.23), (1.24). By force of (34) we obtain for the holomorphic function $\varphi(z)$ the boundary conditions:

1) $\text{Re}\left[\varphi^+(x)\right] = f(x)$,  $\text{Im}\left[\varphi^-(x)\right] = g(x)$,  $x \in L$,
2) $\text{Re}\left[\varphi^\pm(x)\right] = f_1(x)$,  $x \in L_1$   $\text{Im}\left[\varphi^\pm(x)\right] = g_1(x)$,  $x \in L_2$.

Thus $\varphi(z)$ is defined by these conditions (Problems 1, 2).

To solve the problems considered below we will need the Riemann–Schwartz symmetric principle of reflection in elasticity [Ob4]: Let $D$ be some domain in the half plane $y > 0$ the boundary of which consists of segments $L$ of the $x$ axis. $D$ can be the half plane too. Suppose that on $L$ we have the cases:

(1) the normal displacement $v(x, y)$ and the tangent stress $X_y(x, y)$ are zero, or
(2) the normal stress $Y_y(x, y)$ and the tangent displacement $u(x, y)$ are zero.
Then in case (1) the functions

$$u(x, y), v(x, y) = \begin{cases} u(x, y), v(x, y) \\ \quad \text{in } D, \\ u(x, -y), -v(x, -y) \\ \quad \text{in } D^*, \end{cases}$$

$$Y_y(x, y), X_y(x, y) = \begin{cases} Y_y(x, y), X_y(x, y) \\ \quad \text{in } D, \\ Y_y(x, -y), -X_y(x, -y) \\ \quad \text{in } D^*, \end{cases} \tag{1.27}$$

or in case (2) the functions

$$u(x, y), v(x, y) = \begin{cases} u(x, y), v(x, y) \\ \quad \text{in } D, \\ -u(x, -y), v(x, -y) \\ \quad \text{in } D^*, \end{cases}$$

$$Y_y(x, y), X_y(x, y) = \begin{cases} Y_y(x, y), X_y(x, y) \\ \quad \text{in } D, \\ -Y_y(x, -y), X_y(x, -y) \\ \quad \text{in } D^*, \end{cases} \quad (1.28)$$

are the solution of the elasticity equations in $D \cup D^* \cup L$.

Let $D$ be the half-circular domain $|z| < 1$, $y > 0$, $D^+$ be the half plane $x > 0$ with cuts along the segments $[a_k, b_k]$, $k = 1, \ldots, n$, of the $x$ axis, and $D_1$ be the half plane $x > 0$ with cuts along the circular arcs $|z| = 1$, $x > 0$.

**Problem 11.** Find the equilibrium of $D$ by the conditions

$$v(x, 0) = 0, \quad X_y(x, 0) = 0, \quad |x| < 1. \quad (1.29)$$

On $|z| = 1$, $y > 0$ one has the boundary conditions considered above for the circular domain.

By force of (1.27) this problem can be reduced to the problem for the circular domain $|z| < 1$ with corresponding conditions. In particular, the conditions on $|z| = 1$, $y < 0$ are defined by the condition given on $|z| = 1$, $y > 0$. The solution is constructed as above for the circular domain.

**Problem 12.** Find the equilibrium of $D^+$ when the displacements vanish at infinity,

$$v(0, y) = 0, \quad X_y(0, y) = 0, \quad -\infty < y < \infty,$$

and on the segments of the $x$ axis, one has the boundary conditions considered above for the plane with cuts along the $x$ axis. In the case of domain $D_1$, analogous conditions are considered.

Again using the Riemann–Schwartz symmetric principle, this problem can be reduced to the corresponding problem for the plane with cuts along segments of the $x$ axis or along arcs of the circle $|z| = 1$. Thus all these problems can be solved in quadratures. It is understood that if in place of (1.29) one has $u(x, 0) = 0$, $Y_y(x, 0) = 0$, $|x| < 1$, then (1.28) would be used.

## 2 BVP for Generalized Holomorphic Functions

To solve BVP for generalized holomorphic functions in quadratures one must consider the equation with constant coefficients:

$$Lw \equiv \frac{\partial w}{\partial \bar{z}} + B\bar{w} = 0, \quad w = u - iv. \tag{2.1}$$

If $w$ is the solution of this equation, then it is also the solution of the Helmholtz equation:

$$\Delta w - 4|B|^2 w = 0. \tag{2.2}$$

It is easy to obtain

$$w_2 L w_1 = \frac{\partial w_1 w_2}{\partial \bar{z}} - w_1 \left( \frac{\partial w_2}{\partial \bar{z}} - \overline{B w_2} \right) + B w_2 \bar{w}_1 - \overline{B w_2} w_1. \tag{2.3}$$

The operator

$$L^* w \equiv \frac{\partial w}{\partial \bar{z}} - \overline{B w} \tag{2.4}$$

is the adjoint of $Lw$. Thus one has

$$w_2 L w_1 + w_1 L^* w_2 = \frac{\partial w_1 w_2}{\partial \bar{z}} + B w_2 \bar{w}_1 - \overline{B w_2} w_1. \tag{2.5}$$

By force of (2.5) we obtain for any regular functions $w_1, w_2$

$$\operatorname{Re} 2 \iint_D [w_2 L w_1 + w_1 L^* w_2] dx\, dy = -\operatorname{Re} \left[ i \int_\Gamma w_1 w_2\, dt \right], \tag{2.6}$$

where regular functions have continuous first-order derivatives in $D$ and are continuous on $D \cup \Gamma$.

As is well known, the real fundamental solution of equation (2.2) is

$$g(r) = -i H_0^{(1)}(i 2|B| r), \tag{2.7}$$

where $H_0^{(1)}$ is the zeroth-order Hankel function with asymptotic behavior

# I. Two-Dimensional Cases

$$H_0^{(1)}(i2|B|r) = \frac{2i}{\pi} \ln r + \cdots, \quad \text{as } r \to 0, \quad r = |z|, \tag{2.8}$$

where the dots indicate the finite part for $r \to 0$ and

$$H_0^{(1)}(i2|B|r) = -i\sqrt{\frac{2}{\pi r}}[1 + O(r^{-1})]e^{-2r|B|} \quad \text{for } r \to \infty. \tag{2.9}$$

It is easy to show that equation $L^*w = 0$ has two linearly independent fundamental solutions

$$\overset{1}{w}(z) = \frac{\partial g}{\partial z} + \overline{B}g, \quad \overset{2}{w}(z) = i\left(\frac{\partial g}{\partial z} - \overline{B}g\right). \tag{2.10}$$

Using (2.8) for any regular function $w(z)$ one can prove

$$\lim_{\varepsilon \to 0} \int_{C_\varepsilon} w(t)\overset{k}{w}(t-z)\,dt = 2(i)^k w(z), \quad k = 1, 2, \tag{2.11}$$

where $C_\varepsilon$ is a circle of radius $\varepsilon$ with a center at the point $z$. Using (2.10) it follows immediately from the obvious equalities

$$\frac{\partial \ln|z|}{\partial z} = \frac{1}{2z}, \quad \lim_{\varepsilon \to 0} \int_{C_\varepsilon} w(t) \ln|t - z|\,dt = 0,$$

$$\lim_{\varepsilon \to 0} \int_{C_\varepsilon} w(t) \frac{\partial \ln|t-z|}{\partial t}\,dt = \lim_{\varepsilon \to 0} \frac{i}{2} \int_0^{2\pi} w(z + \varepsilon e^{i\vartheta})\,d\vartheta = i\pi w(z).$$

Now let $w$ be the regular solution of equation (2.1). Then by force of (2.6) we have

$$\operatorname{Re} i \int_\Gamma w \frac{\partial g}{\partial t}\,dt = \lim_{\varepsilon \to 0} \operatorname{Re} i \int_{\Gamma_\varepsilon} w \overset{1}{w}\,dt = \lim_{\varepsilon \to 0} \operatorname{Re} \frac{i}{\pi} \int_{\Gamma_\varepsilon} \frac{w(t)\,dt}{t - z} = -2\operatorname{Re} w(z),$$

$$\operatorname{Re} \int_\Gamma w \frac{\partial g}{\partial t}\,dt = \lim_{\varepsilon \to 0} \operatorname{Re} \frac{1}{\pi} \int_{\Gamma_\varepsilon} \frac{w(t)\,dt}{t-z} = 2\operatorname{Re} iw(z).$$

These can be written as

$$\operatorname{Im} \int_\Gamma w \frac{\partial g}{\partial t}\,dt = 2\operatorname{Re} w(z),$$

$$\operatorname{Re} \int_\Gamma w \frac{\partial g}{\partial t}\,dt = -2\operatorname{Im} w(z).$$

From these equalities one obtains

$$\int_\Gamma w \frac{\partial g(t-z)}{\partial t} dt = 2i w(z), \quad z \in D. \tag{2.12}$$

Thus by force of (2.10), (2.12) we have

$$w(z) = \frac{1}{2i} \int_\Gamma \left[ w(t) \frac{\partial g(t-z)}{\partial t} dt - Bg(t-z)\overline{w(t)}\, d\bar{t} \right]. \tag{2.13}$$

By virtue of (2.7), (2.8) one has

$$\lim_{B \to 0} \frac{\partial g(t-z)}{\partial t} = \frac{1}{\pi(t-z)}. \tag{2.14}$$

That is why for $B \to 0$ equality (2.13) becomes the Cauchy integral formula and is called the generalized Cauchy integral formula.

Let $\varphi(t)$ be a function given on $\Gamma$ and consider the generalized Cauchy-type integral

$$w(z) = \frac{1}{2i} \int_\Gamma \left[ \varphi(t) \frac{\partial g(t-z)}{\partial t} dt - Bg(t-z)\overline{\varphi(t)}\, d\bar{t} \right]. \tag{2.15}$$

Obviously, (2.15) exists for every $z$ not in $\Gamma$. One can verify that $w(z)$ is the solution of (2.1) in $D$ and $D^-$, $w(\infty) = 0$. Besides, if $\varphi(t)$ is Hölder-continuous on $\Gamma$, by analogy with the Plemelj–Sokhotzki formula from (2.15) one can get

$$w^+(t) - w^-(t) = \varphi(t), \quad t \in \Gamma. \tag{2.16}$$

Let $\Gamma_1, \Gamma_2, \ldots, \Gamma_n$ be smooth lines situated inside the domain $D$ with the boundary $\Gamma$.

**Problem 1.** Find a piecewise generalized holomorphic function that is a solution of (2.1) and that vanishes at infinity by condition (2.16).

Note that this problem is investigated in the sense of solvability for a variable $B$ [Ve]. It is clear that for constant $B$ the solution is represented in quadratures by (2.15).

**Problem 2.** Find in the half plane $\operatorname{Im} z > 0$ a solution of (2.1) that vanishes at infinity by the condition

$$\operatorname{Re}[w(t)] = f(t), \quad t \in R(-\infty, \infty), \tag{2.17}$$

where $f(t)$ is a given Hölder-continuous function.

**Solution.** To solve this problem we need some properties of the equation (2.1). Let $L$ be a straight line $\alpha x + \beta y = 0$. Without loss of generality one can assume $\alpha^2 + \beta^2 = 1$. The symmetric points with respect to $L$ are $z$ and

$$z^* = -(\alpha + i\beta)^2 \bar{z}. \tag{2.18}$$

It is well known that if $w(z)$ is a holomorphic function, then $\psi(z) = \overline{w(z^*)}$ is also holomorphic and the symmetric principle of reflection formulated above is again justified. The question is, do generalized holomorphic functions have that property?

For (2.1) with $B = a + ib$ when the boundary of $D$ contains an arc of a circle or a segment of the straight line $L$, where $(\alpha, \beta)$ and $(a, b)$ are independent, the above property is not justified. But it is possible to define $(a, b)$ by $(\alpha, \beta)$ or vice versa in such a way that the Riemann–Schwartz principle of reflection is just. By (2.18) equation (2.1) is transformed into

$$\frac{\partial \overline{w(z^*)}}{\partial \bar{z}^*} - \overline{B}(\alpha + i\beta)^2 \overline{w(z^*)} = 0. \tag{2.19}$$

If

$$\overline{B(z^*)}(\alpha + i\beta)^2 + B(z) = 0, \tag{2.20}$$

then for the function $\Omega(z) = \overline{w(z^*)}$ one has the same equation (2.1). With the conditions as for holomorphic functions, the Riemann–Schwartz symmetric principle of reflection is justified. For instance, if $L$ is a segment of the axis $y = 0$, then $z^* = \bar{z}$ and $\overline{B(\bar{z})} - B(z) = 0$. Thus if $B$ is constant, then it must be real. If $L$ is a segment of $x = 0$, then $z^* = -\bar{z}$ and $B(z) + \overline{B(-\bar{z})} = 0$; i.e., if $B$ is constant, its real part must be zero.

Thus for holomorphic functions, the class of domains for which the symmetric principle of reflection is justified is wider than for generalized holomorphic functions. That is why many BVP that are solved in quadratures for holomorphic functions are impossible to solve for generalized holomorphic functions.

To solve Problem 2 consider the function

$$\Omega(z) = \begin{cases} w(z), & y > 0, \\ -\overline{w(\bar{z})}, & y < 0. \end{cases} \tag{2.21}$$

It is a solution of (2.1) if $B$ is a real constant. Then the boundary condition (2.17) can be written as

$$\Omega^+(t) - \Omega^-(t) = 2f(t), \quad t \in R.$$

By virtue of (2.15), (2.16) the solution of Problem 2 can be represented in the form

$$w(z) = -i \int_R \left[ \frac{\partial g(t-z)}{\partial t} - Bg(t-z) \right] f(t)\, dt. \tag{2.22}$$

**Problem 3 (Compound BVP).** Let $D$ be the half plane $y > 0$. $L_1, L_2, \ldots, L_n$ are smooth lines situated in $D$. Find the solution of (2.1) in $D$ that vanishes at infinity with the jump line $L = \bigcup_1^n L_k$ by the boundary conditions

$$\text{Re}[w(t)] = f(t), \quad t \in R, \tag{2.23}$$
$$w^+ - w^- = \varphi(t), \quad t \in L. \tag{2.24}$$

**Solution.** The solution of this problem is represented in the form

$$w(z) = w_1(z) + w_2(z).$$

Function $w_1(z)$ is defined by condition (2.24), i.e., by formula (2.15). Function $w_2(z)$ is a generalized holomorphic function in the half plane $y > 0$ and, by force of (2.23), is defined by the condition

$$\text{Re}[w_2(t)] = f(t) - \text{Re}[w_1(t)] \equiv F(t), \quad t \in R.$$

Thus $w_2(z)$ is defined as (2.22).

Let $D$ be the $z$ plane with cuts along the segments $L_k$ ($k = 1, \ldots, n$) of the $x$ axis. In this case BVP for generalized holomorphic functions cannot be solved as successfully as for holomorphic functions. It is possible to define a generalized holomorphic function in $D$ that vanishes at infinity with $B$ a real constant only by the conditions

$$\text{Re}\, w^\pm(x) = \pm f(x). \tag{2.25}$$

By force of (2.15), (2.22) its representation is

$$w(z) = -i \int_L \left[ \frac{\partial g(t-z)}{\partial t} - Bg(t-z) \right] f(t)\, dt, \quad L = \bigcup_1^n L_k.$$

But if in place of conditions (2.25) one has

$$\operatorname{Re}[w^{\pm}(x)] = f(x), \quad x \in L, \tag{2.26}$$

then $w(z)$ cannot be defined in quadratures. It is possible if $L$ is only one segment $[0, \infty]$. We will solve it below.

Consider the following problem in the half plane $y > 0$ with the conditions

$$\operatorname{Re}[w(x)] = f(x), \quad x > 0, \quad \operatorname{Im}[w(x)] = 0, \quad x < 0. \tag{2.27}$$

Using the Riemann–Schwartz principle of reflection these can be reduced to conditions (2.26). For the quarter plane ($x > 0$, $y > 0$) it is not possible to solve effectively any problem for generalized holomorphic functions. But it is possible for the Helmholtz equation (2.2).

**Problem 4.** Find $u(x, y)$ for $x > 0$, $y > 0$, vanishing at infinity with the conditions

$$\Delta u - k^2 u = 0, \quad x > 0, \quad y > 0,$$
$$u(x, 0) = f(x), \quad x > 0,$$
$$u(0, y) = 0, \quad y > 0, \quad \text{or} \quad \left.\frac{\partial u}{\partial x}\right|_{x=0} = 0, \quad y > 0.$$

**Solution.** Using Riemann–Schwartz principle this problem can be easily reduced to the corresponding problem in the half plane $y > 0$ with the conditions

$$u(x, 0) = f(x), \quad x > 0, \quad u(x, 0) = -f(-x), \quad x < 0, \tag{2.28}$$

or

$$u(x, 0) = f(x), \quad x > 0, \quad u(x, 0) = f(-x), \quad x < 0,$$

By force of (2.22) the solution can be represented as

$$u(x, y) = -\operatorname{Re} \int_R u(t, 0) \frac{\partial H_0^{(1)}(ikr)}{\partial t} dt, \quad r^2 = (x-t)^2 + y^2. \tag{2.29}$$

## 3  Mixed BVP for Generalized Holomorphic Functions

Let $D_1$ be the half plane $y > 0$, where $z = x + iy$. Let $w(z) = u - iv$ be the solution of the equation

## 3. Mixed BVP for Generalized Holomorphic Functions

$$\frac{\partial w}{\partial \bar{z}} + B\bar{w} = 0, \qquad (3.1)$$

where $B$ is a real constant.

**Problem 1.** Find in $D_+$ the regular solution of (3.1) that vanishes at infinity by the conditions

$$u(x, 0) = \varphi(x), \quad x > 0, \quad v(x, 0) = \psi(x), \quad x < 0, \qquad (3.2)$$

where the given functions $\varphi, \psi$ are of class $L$ and vanish at infinity.

**Solution.** This problem can be solved using Fourier Integral Transforms (FIT). A short version of the theory is given in the introduction of the next chapter. The system (3.1) can be written in the form

$$\begin{aligned} \frac{\partial u}{\partial x} + \frac{\partial v}{\partial y} + Bu &= 0, \\ \frac{\partial u}{\partial y} - \frac{\partial v}{\partial x} + Bv &= 0. \end{aligned} \qquad (3.3)$$

The FIT of a function $L$ with respect to the variable $x$ is, by definition,

$$\widehat{u} = \frac{1}{\sqrt{2\pi}} \int_R u(x, y) e^{-ixt} dx.$$

It is easy to obtain

$$\left[\widehat{\frac{\partial u}{\partial x}}\right] = it\widehat{u}.$$

The inverse formula for $u \in L, \widehat{u} \in L$ is

$$u(x, y) = \frac{1}{\sqrt{2\pi}} \int_R \widehat{u}(t, y) e^{ixt} dt. \qquad (3.4)$$

Then one can obtain by the FIT of (3.3)

$$\begin{aligned} \frac{d\widehat{v}}{dy} + (it + B)\widehat{u} &= 0, \\ \frac{d\widehat{u}}{dy} - (it - B)\widehat{v} &= 0, \end{aligned}$$

34     I. Two-Dimensional Cases

which represents the solution of these equations as

$$\hat{u} = A_1(t)e^{\lambda y}, \quad \hat{v} = A_2(t)e^{\lambda y}. \tag{3.5}$$

By force of (3.4), (3.5) for $A_1$, $A_2$ one has

$$A_2\lambda + (it + B)A_1 = 0,$$
$$-A_2(it - B) + \lambda A_1 = 0.$$

These equations have nonzero solutions for

$$\lambda = \pm\sqrt{t^2 + B^2}$$

and correspondingly $A_2 = -iA_1\frac{t-iB}{\lambda}$. Thus by the inverse formula from (3.5) we have

$$u(x, y) = \frac{1}{\sqrt{2\pi}} \int_R A_1(t) e^{-\sqrt{t^2+B^2}\,y} e^{ixt} dt,$$
$$v(x, y) = \frac{i}{\sqrt{2\pi}} \int_R A_1(t) \sqrt{\frac{t-iB}{t+iB}} e^{-\sqrt{t^2+B^2}\,y} e^{ixt} dt. \tag{3.6}$$

To define $A_1(t)$ by boundary conditions (3.2) we have the coupled integral equations

$$\frac{1}{\sqrt{2\pi}} \int_R A_1(t) e^{ixt} dt = f(x), \quad x > 0,$$
$$\frac{i}{\sqrt{2\pi}} \int_R A_1(t) \sqrt{\frac{t-iB}{t+iB}} e^{ixt} dt = \psi(x), \quad x < 0, \tag{3.7}$$

which are solved using the Wiener–Hopf method. In the first equation in place of $x$ write $x + \xi$, $\xi > 0$, and in the second equation in place of $x$ write $x - \xi$, $\xi > 0$. Then multiply the first equasstion by the unknown function $N_1(\xi)$ and the second equation by the function $N_2(\xi)$. Integrate by $\xi$ on $[0, \infty]$. Supposing a change of the order of integration is possible one gets

$$\frac{1}{\sqrt{2\pi}} \int_R A_1(t) e^{ixt} dt \int_0^\infty N_1(\xi) e^{i\xi t} d\xi =$$
$$= \int_0^\infty N_1(\xi) f(x+\xi) d\xi \equiv f_1(x),$$
$$\frac{1}{\sqrt{2\pi}} \int_R A_1(t) \sqrt{\frac{t-iB}{t+iB}} e^{ixt} dt \int_0^\infty N_2(\xi) e^{-i\xi t} d\xi =$$
$$= -i \int_0^\infty N_2(\xi) \psi(x-\xi) d\xi \equiv g_1(x), \quad x < 0. \tag{3.8}$$

## 3. Mixed BVP for Generalized Holomorphic Functions

Consider two new unknown functions

$$\Phi^+(t) = \int_0^\infty N_1(\xi) e^{i\xi t} d\xi, \quad \Phi^-(t) = \int_0^\infty N_2(\xi) e^{-i\xi t} d\xi. \tag{3.9}$$

It is easy to see that if in place of $t$ one writes the complex variable $z = t + i\sigma$, $\Phi^+(t)$ will be the limit as $\sigma \to 0$ of a holomorphic function of $z$ in the half plane $\sigma > 0$ and $\Phi^-(t)$ will be the limit as $\sigma \to 0$ of a holomorphic function of $z$ in the plane $\sigma < 0$. Thus (3.8) is written as

$$\frac{1}{\sqrt{2\pi}} \int_R A_1(t) \Phi^+(t) e^{ixt} dt = f_1(x), \quad x > 0,$$

$$\frac{1}{\sqrt{2\pi}} \int_R A_1(t) \sqrt{\frac{t-iB}{t+iB}} \Phi^-(t) e^{ixt} dt = g_1(x), \quad x < 0, \quad B > 0. \tag{3.10}$$

Let the holomorphic function $\Phi(z)$ that vanishes at infinity satisfy the condition

$$\Phi^+(t) = \sqrt{\frac{t-iB}{t+iB}} \Phi^-(t), \quad t \in R.$$

It is clear that

$$\Phi^+(z) = \frac{1}{\sqrt{z+iB}}, \quad \sigma > 0, \quad \Phi^-(z) = \frac{1}{\sqrt{z-iB}}, \quad \sigma < 0.$$

By force of (3.9) and using the inverse formula one has

$$N_1(\xi) = \frac{1}{2\pi} \int_R \frac{1}{\sqrt{t+iB}} e^{-i\xi t} dt, \quad N_2(\xi) = \frac{1}{2\pi} \int_R \frac{1}{\sqrt{t-iB}} e^{i\xi t} dt, \quad \xi > 0.$$

These integrals can be calculated in the following way. $\Phi^+(z)$ is a holomorphic function for $\sigma > 0$, and $e^{-i\xi t}$ for $t = \rho(\cos\alpha + i\sin\alpha)$, $0 < \alpha < \pi$, tends to $\infty$ as $\rho \to \infty$. But $\Phi^+(z)$ for $\sigma < 0$ has a singularity at $t = -iB$, and $e^{-i\xi t}$ tends to zero as $\rho \to \infty$, $\pi < \alpha < 2\pi$. $\Phi^+(z)$ is a multivalued function in the half plane $\sigma < 0$. Therefore, to use Cauchy's theorem we consider a domain where $\Phi^+(z)$ is single valued—the half plane $\sigma < 0$ with $t = 0$, $-\infty < \sigma < -B$. On the left side of this line let $\Phi^+(i\sigma) = \frac{-1}{\sqrt{i\sigma+iB}}$ and on the right-hand side $\Phi^+(i\sigma) = \frac{1}{\sqrt{i\sigma+iB}}$. Thus Cauchy's formula gives

$$N_1(\xi) = \frac{1}{2\pi} \int_R \frac{e^{-i\xi t} dt}{\sqrt{t+iB}} = \frac{1}{\pi} \int_{-\infty}^{-iB} \frac{e^{-i\xi z}}{\sqrt{z+iB}} dz.$$

36   I. Two-Dimensional Cases

Let $z + iB = -i\tau$, then

$$N_1(\xi) = \frac{1}{\sqrt{i\pi}} \int_0^\infty \frac{e^{-\xi B} e^{-\xi \tau}}{\sqrt{\tau}} d\tau = \frac{2e^{-\xi B - i\frac{\pi}{4}}}{\pi} \int_0^\infty e^{-\xi \tau^2} d\tau =$$

$$= \frac{1}{\sqrt{\pi}\sqrt{\xi}} \exp\left[-\xi B - i\frac{\pi}{4}\right], \quad \xi > 0. \tag{3.11}$$

If we consider $\Phi^-(z)$ in the half plane $\sigma > 0$, we similarly obtain

$$N_2(\xi) = \frac{1}{\sqrt{\pi}\sqrt{\xi}} \exp\left[-\xi B + i\frac{\pi}{4}\right], \quad \xi > 0. \tag{3.12}$$

By force of (3.8) we can write

$$\frac{1}{\sqrt{2\pi}} \int_R A_1(t) \frac{1}{\sqrt{t+iB}} e^{ixt} dt \equiv F(x) \equiv \begin{cases} f_1(x), & x > 0, \\ g_1(x), & x < 0. \end{cases}$$

Using the inverse formula, $A_1(t)$ is defined

$$A_1(t) = \frac{\sqrt{t+iB}}{\sqrt{2\pi}} \int_R F(x) e^{-ixt} dx,$$

and the solution of problem (3.2) is represented in the form (3.6).

Let $D$ be the plane $z = x + iy$ with a cut along the half axis $y = 0, x > 0$.

**Problem 2.** Find the regular solution of (3.1) in $D$ that vanishes at infinity by the conditions

$$u^\pm(x, 0) = f^\pm(x), \quad x > 0. \tag{3.13}$$

**Solution.** The solution of this problem can be reduced to the solution of the problem (3.2). The solution can be represented in the form

$$w = w_1 + w_2,$$

where

$$w_1 = \frac{w(x, y) + w(x, -y)}{2}, \quad w_2 = \frac{w(x, y) - w(x, -y)}{2} \tag{3.14}$$

## 3. Mixed BVP for Generalized Holomorphic Functions

provide the solution of (3.1) and by force of (3.13) satisfy the conditions

$$\operatorname{Re} w_1^\pm = \frac{f^+ + f^-}{2} \equiv f(x), \quad \operatorname{Re} w_2^\pm = \pm \frac{f^+ - f^-}{2} = \pm \varphi(x), \quad x > 0, \ y = 0. \tag{3.15}$$

By force of (2.25) $w_2$ is represented as

$$w_2(z) = -i \int_0^\infty \left[ \frac{\partial g(t-z)}{\partial t} - Bg(t-z) \right] \varphi(t) \, dt.$$

To determine $w_1(z) = u + iv$ we consider the problem for the half plane $y > 0$ with the conditions

$$v(x, 0) = 0, \quad x < 0, \quad u(x, 0) = f(x), \quad x > 0.$$

The solution of this problem is given in the form (3.6). Then using the symmetric principle of reflection we obtain $w_1(z)$ satisfying the condition (3.15).

Now consider in $D_+$ the bigeneralized holomorphic function $w(z)$

$$\frac{\partial w}{\partial \bar{z}} + B\bar{w} = F(z), \tag{3.16}$$

$$\frac{\partial F}{\partial \bar{z}} + B\bar{F} = 0, \tag{3.17}$$

where $B$ is a real constant.

**Mixed BVP.** Find the regular solution of (3.16), (3.17) in $D_+$ that vanishes at infinity by the conditions

$$\operatorname{Re} w(x, 0) = \varphi_1(x), \quad \operatorname{Re} \left. \frac{\partial w}{\partial \bar{z}} \right|_{y=0} = \varphi_2(x), \quad x > 0, \tag{3.18}$$

$$\operatorname{Im} w(x, 0) = \varphi_3(x), \quad \operatorname{Im} \left. \frac{\partial w}{\partial \bar{z}} \right|_{y=0} = \varphi_4(x), \quad x < 0, \tag{3.19}$$

where the given functions $\varphi_k(x)$ vanish at infinity.

**Solution.** It is clear that by these conditions and (3.16) one can define

$$\operatorname{Re} F(x, 0) = \varphi(x), \quad x > 0, \quad \text{and} \quad \operatorname{Im} F(x, 0) = \psi(x), \quad x < 0.$$

Thus for $F(z)$ we have problem (3.2), the solution of which is given above. Then the solution of (3.16) can be represented as

$$w(z) = w_1(z) + w_2(z), \tag{3.20}$$

where $w_1(z)$ is the solution of the homogeneous equation (3.1) and $w_2(z)$ is the partial solution of the nonhomogeneous equation. By conditions (3.18), (3.19), $w(z)$ is defined in quadratures.

For the domain $D$ with a cut along the half axis $y = 0$, $x > 0$ the problem can be solved in an analogous way.

**Problem 3.** Find the regular solution of (3.16), (3.17) in $D$ that vanishes at infinity with conditions

$$\mathrm{Re}[w^\pm(x, 0)] = f^\pm(x),$$
$$\mathrm{Re}\left[\frac{\partial w}{\partial \bar z}\right]^\pm_{y=0} = \varphi^\pm(x).$$

Using problem (3.13), $w(z)$ can be defined in quadratures with these conditions too.

## 4  BVP for Beltrami and Generalized Beltrami Equations

One of the important generalizations of the Cauchy–Riemann system is the well-known Beltrami equation

$$\frac{\partial w}{\partial \bar z} + q \frac{\partial w}{\partial z} = 0, \quad w = u - iv, \quad |q| < 1. \tag{4.1}$$

Its several generalizations have been investigated. One of them is

$$\frac{\partial w}{\partial \bar z} + q_1 \frac{\partial w}{\partial z} + q_2(z) \frac{\partial \overline{w}}{\partial \bar z} = 0. \tag{4.2}$$

To classify this equation consider the curve [Ob2]

$$\Gamma : \varphi(x, y) = 0, \quad \frac{\partial \varphi}{\partial y} \neq 0. \tag{4.3}$$

This curve is characteristic for (4.2) if the solution

$$w(z) = f(x, y), \tag{4.4}$$

is given on $\Gamma$. One cannot define the first-order derivatives of $w$ by equation (4.2). By the equalities on $\Gamma$

$$\frac{\partial \varphi}{\partial x} + \frac{\partial \varphi}{\partial y}\frac{dy}{dx} = 0, \quad \frac{\partial w}{\partial x} = \frac{\partial w}{\partial y}\frac{\tau_0}{\tau_1} + F,$$

where $F$ is a given function on $\Gamma$ defined by (4.4). Let $\tau_0 = \frac{\partial \varphi}{\partial x}$, $\tau_1 = \frac{\partial \varphi}{\partial y}$. By the above equalities, equation (4.2) gives on $\Gamma$ equation

$$(\tau + q_1\overline{\tau})\frac{\partial w}{\partial y} + q_2\tau\frac{\partial \overline{w}}{\partial y} = F_1, \quad \tau = \tau_0 + i\tau_1, \tag{4.5}$$

and its conjugate

$$\overline{q_2}\overline{\tau}\frac{\partial w}{\partial y} + (\overline{\tau} + \overline{q_1}\tau)\frac{\partial \overline{w}}{\partial y} = \overline{F}_1. \tag{4.6}$$

These equations are uniquely solvable for $\dfrac{\partial w}{\partial y}, \dfrac{\partial \overline{w}}{\partial y}$ if

$$Q(\tau) \equiv |\tau + q_1\overline{\tau}|^2 - |q_2|^2|\tau|^2 \neq 0.$$

Let $q_1 = a + ib$, then

$$Q(\tau) = [1 + |q_1|^2 - |q_2|^2 + 2a]\tau_0^2 + 4b\tau_0\tau_1 + (1 + |q_1|^2 - |q_2|^2 - 2a)\tau_1^2.$$

If $Q(\tau) = 0$ the above derivatives in (4.5), (4.6) can not be determined for arbitrary $F$. The associated discriminant for $Q$ is

$$M(q_1, q_2) = 4b^2 - \left[(1 + |q_1|^2 - |q_2|^2)^2 - 4a^2\right], \quad \text{i.e.,}$$
$$M(q_1, q_2) = \left[(1 + |q_1|)^2 - |q_2|^2\right]\left[|q_2|^2 - (1 - |q_1|)^2\right] =$$
$$= \left[(|q_1| + |q_2|)^2 - 1\right]\left[1 - (|q_1| - |q_2|)^2\right]. \tag{4.7}$$

Thus equation (4.2) is elliptic when $Q(\tau)$ is of definite quadratic form with respect to $(\tau_0, \tau_1)$, i.e., $M < 0$. If it is an indefinite form ($M > 0$) then (4.2) is hyperbolic. If it is a degenerate form ($M = 0$) then (4.2) is parabolic.

Hence (4.2) is elliptic if $q_1, q_2$ satisfy one of the conditions:
(a) $|q_1| + |q_2| < 1$ and $||q_1| - |q_2|| < 1$, or equivalently, $||q_1| - |q_2|| \leq |q_1| + |q_2|$,

$$|q_1| + |q_2| < 1. \tag{4.8}$$

40    I. Two-Dimensional Cases

(b) $|q_1| + |q_2| > 1$, $||q_1| - |q_2|| > 1$, or equivalently,

$$\left||q_1| - |q_2|\right| > 1. \tag{4.9}$$

Equation (4.2) is hyperbolic if

$$\left||q_1| - |q_2|\right| < 1 \quad \text{and} \quad |q_1| + |q_2| > 1, \tag{4.10}$$

and is parabolic if

$$|q_1| + |q_2| = 1 \quad \text{or} \quad \left||q_1| - |q_2|\right| = 1. \tag{4.11}$$

Now we prove that equation (4.2) with condition (4.9) can be reduced to equation (4.2) with condition (4.8). Indeed by (4.2) we obtain

$$\bar{q}_2 \frac{\partial \bar{w}}{\partial \bar{z}} + q_1 \bar{q}_2 \frac{\partial w}{\partial z} + |q_2|^2 \frac{\partial \bar{w}}{\partial z} = 0,$$

$$q_1 \frac{\partial \bar{w}}{\partial z} + |q_1|^2 \frac{\partial \bar{w}}{\partial \bar{z}} + q_1 \bar{q}_2 \frac{\partial w}{\partial z} = 0.$$

Subtraction of these equations yields

$$\frac{\partial \bar{w}}{\partial \bar{z}} + \frac{q_1}{|q_1|^2 - |q_2|^2} \frac{\partial \bar{w}}{\partial z} - \frac{\bar{q}_2}{|q_1|^2 - |q_2|^2} \frac{\partial w}{\partial \bar{z}} = 0. \tag{4.12}$$

It is easy to check that if $q_1, q_2$ satisfy condition (4.9), then the coefficients of this equation with respect to $\bar{w}$ satisfy (4.8). Thus, in the case of ellipticity we will consider equation (4.2) only with conditions (4.8).

If $q_2 = 0$ the Beltrami equation (4.1) can be elliptic for $|q_1| \neq 1$ and parabolic for $|q_1| = 1$. In the elliptic case it is sufficient to consider

$$|q| < 1. \tag{4.13}$$

Because we are interested in solving problems explicitly in quadratures, (4.1), (4.2) are considered with constant coefficients.

We also consider the equation with constant coefficients

$$\frac{\partial w}{\partial \bar{z}} + q \frac{\partial w}{\partial z} + B \bar{w} = 0, \quad |q| < 1. \tag{4.14}$$

## 4. BVP for Beltrami and Generalized Beltrami Equations

Consider the linear transformation

$$\zeta = \frac{z - q\bar{z}}{1 - |q|^2} \tag{4.15}$$

Then equation (4.14) is reduced to

$$\frac{\partial w}{\partial \bar{\zeta}} + B\bar{w} = 0, \quad \zeta = \xi + i\eta, \tag{4.16}$$

i.e., $w = w(\zeta)$ as function of $\zeta$ is a generalized holomorphic function and, if $B = 0$, is a holomorphic function. If $q = |q|e^{i\alpha}$, (4.1) can be reduced to the form

$$\frac{\partial w}{\partial \bar{z}_1} + |q|\frac{\partial w}{\partial z_1} = 0, \quad z_1 = ze^{-\frac{i\alpha}{2}}, \tag{4.17}$$

$$z_1 = x_1 + iy_1.$$

That is why the transformation (4.15) can be considered with real $q$. The straight line $y = 0$ is transformed into $\eta = 0$, and symmetric points with respect to $y = 0$ in the $z$ plane is transformed into symmetric points with respect to $\eta = 0$ in the $\zeta$ plane.

From the Riemann–Schwartz principle of reflection for holomorphic functions one has: if $w(z)$ is a regular solution of (4.1) in $D$ the boundary $\Gamma$ of which contains part of the straight line $y = 0$ and

$$\text{Re } w(x) = 0, \quad x \in \Gamma,$$

then

$$w_1(z) = \begin{cases} w(z), & z \in D, \\ -\overline{w(\bar{z})}, & z \in D^*, \end{cases} \tag{4.18}$$

is a regular solution of (4.1) in the domain $D \cup D^* \cup \Gamma$. But if $q$ is not real, by force of (4.17) and (4.15), symmetric points with respect to the straight line $y = x \, \text{tg}\, \alpha/2$ in the $z$ plane are transformed into symmetric points with respect to $\eta = 0$ in the $\zeta$ plane. Thus, for (4.1) the straight line for which the Riemann–Schwartz principle of reflection is true is defined uniquely.

Now let $D$ be the domain the boundary of which contains an arc of the circle $|z| = 1$. Unlike holomorphic functions if $w(z)$ is a solution of (4.1), then

$$w_1(z) = \overline{w\left(\frac{1}{\bar{z}}\right)},$$

is not. But if instead of a circle an ellipse is considered, which will be defined properly, the Riemann–Schwartz principle is formulated correspondingly.

Consider the ellipse $\Gamma$

$$\frac{x^2}{a^2} + \frac{y^2}{b^2} = 1. \tag{4.19}$$

Two points $z$, $z^*$ are called symmetric with respect to $\Gamma$ if

$$|z||z^*| = |z_1|^2, \tag{4.20}$$

where $z$, $z^*$ are points lying on a ray through the origin $z = 0$ and $z_1$ is the intersection of this ray with the ellipse. Using (4.20) one can obtain

$$z^* = \frac{z}{|A_1 z - A_2 \bar{z}|^2}, \quad A_1 + A_2 = \frac{1}{b}, \quad A_1 - A_2 = \frac{1}{a}. \tag{4.21}$$

The following statements hold:

(a) By the transformation $\zeta = A_1 z - A_2 \bar{z}$, the domain $D$ inside $\Gamma$ is transformed into the circle $|\zeta| \leq 1$.

(b) If $\zeta$, $\zeta^*$ are symmetric points with respect to $|\zeta| = 1$, then the corresponding points $z$, $z^*$, defined by (4.21), are symmetric with respect to the ellipse $\Gamma$ and vice versa.

(c) For any solution $w(z)$ of (4.1) the function

$$w_1(z) = \overline{w(z^*)},$$

is also a solution if

$$q = \frac{A_2}{A_1}. \tag{4.22}$$

The reflection principle is formulated correspondingly.

If on the arc $\Gamma_1$ of an ellipse one has

$$\operatorname{Re} w(t) = 0, \quad t \in \Gamma_1,$$

then

$$w(z) = \begin{cases} w(z), & z \in D, \\ -\overline{w(z^*)}, & z \in D^*, \end{cases}$$

is the solution of (4.1) in $D \cup D^* \cup \Gamma_1$.

One can see that the reflection principle for the given equation, i.e., for the given $q$, is true for the ellipse defined by (4.21), (4.22) and for an arbitrary ellipse the corresponding equation, i.e., $q$, can be defined by (4.22) and the reflection principle holds again.

These properties of the solutions of (4.1) mean that the BVP solved above for holomorphic functions can also be solved explicitly for the Beltrami equation with real constant $q$.

The BVP solved for generalized holomorphic functions above, by force of (4.14), (4.16), can be solved for equation (4.14) too.

Now consider equation (4.1) when [Ob1]

$$q = \overline{\frac{\varphi'(z)}{\psi'(z)}}, \quad \psi'(z) \neq 0, \tag{4.23}$$

where $\varphi(z)$, $\psi(z)$ are any holomorphic functions of $z$. Then the solution of (4.1) can be represented in the form

$$w = \Phi(\psi(z) - \overline{\varphi(z)}),$$

where $\Phi(\zeta)$ is any holomorphic function of $\zeta = \psi(z) - \overline{\varphi(z)}$. Thus in this case it is understood which BVP can be solved explicitly for equation (4.1). In particular, the domain of $z$ must be transformed into the domain of $\zeta$ so that the BVP for holomorphic functions can be solved explicitly.

Equation (4.2) with complex constant coefficients $q_1 = |q_1|e^{i\alpha}$, $q_2 = |q_2|e^{i\beta}$ can be reduced to the equation (4.2) with real coefficients by the transformation

$$z_1 = ze^{-\frac{i\alpha}{2}}, \quad w_1 = we^{-\frac{i\beta}{2}}.$$

Thus we consider (4.2) where $q_1, q_2$ are real. Its solution can be represented by a holomorphic function of some complex variable, in which case (4.2) is written in the form

$$(1+q_1+q_2)\frac{\partial u}{\partial x}+(1-q_1-q_2)\frac{\partial v}{\partial y}=0,$$
$$(1-q_1+q_2)\frac{\partial u}{\partial y}-(1+q_1-q_2)\frac{\partial v}{\partial x}=0,$$
(4.23$_1$)

From which it is easy to obtain that $u, v$ are solutions of the equation

$$\frac{\partial^2 u}{\partial x^2}\left[(1+q_1)^2-q_2^2\right]+\frac{\partial^2 u}{\partial y^2}\left[(1-q_1)^2-q_2^2\right]=0.$$

Hence by the transformation

$$\xi=\left[(1-q_1)^2-q_2^2\right]^{1/2}x,\quad \eta=\left[(1+q_1)^2-q_2^2\right]^{1/2}y,\quad (4.24)$$

one can see that $u, v$ as functions of $\xi, \eta$ are solutions of the Laplace equation.
By transformation (4.24) system (4.23$_1$) can be written in the form

$$\frac{\partial u}{\partial \xi}\left[(1+q_2)^2-q_1^2\right]^{1/2}+\frac{\partial v}{\partial \eta}\left[(1-q_2)^2-q_1^2\right]^{1/2}=0,$$
$$\frac{\partial u}{\partial \eta}\left[(1+q_2)^2-q_1^2\right]^{1/2}-\frac{\partial v}{\partial \xi}\left[(1-q_2)^2-q_1^2\right]^{1/2}=0.$$
(4.25)

Consider the functions

$$u_1=u\left[(1+q_2)^2-q_1^2\right]^{1/2},$$
$$v_1=v\left[(1-q_2)^2-q_1^2\right]^{1/2}.$$
(4.26)

By the transformation of variables (4.24) and by the function (4.26) equation (4.2) with real $q_1, q_2$ is reduced to the form

$$\frac{\partial w_1}{\partial \bar{\zeta}}=0,\quad w_1=u_1-iv_1,\quad \zeta=\xi+i\eta. \qquad (4.27)$$

Hence $w_1(\zeta)$ is a holomorphic function of $\zeta$. The corresponding function $w=u-iv$ is defined by (4.26). The BVP for (4.2) is reduced to the BVP for holomorphic functions (4.27). The BVP for $w_1$, i.e., for (4.2), can also be solved explicitly in quadratures.

## 5 BVP for Pluriholomorphic, Plurigeneralized Holomorphic, Polyharmonic, Polymetaharmonic Functions and the PluriBeltrami Equation

Consider high-order differential equations obtained by iteration of the Cauchy–Riemann operator $\frac{\partial}{\partial \bar{z}}$

## 5. BVP for Pluriholomorphic ... and the PluriBeltrami Equation

$$\frac{\partial^n w}{\partial \bar{z}^n} = 0 \quad n = 2, 3, \ldots, \tag{5.1}$$

and its generalization

$$P^n w = 0, \quad Pw \equiv \frac{\partial w}{\partial \bar{z}} + B\bar{w}. \tag{5.2}$$

The solutions of (5.1) and (5.2) $w(z) = u(x, y) + iv(x, y)$ are called pluriholomorphic and plurigeneralized holomorphic functions respectively. One can easily see that $w(z)$, the solution of (5.1), is a polyharmonic function too, i.e.,

$$\Delta^n w = 0, \tag{5.3}$$

and the solution of (5.2) with constant $B$ is polymetaharmonic function, i.e.,

$$(\Delta - |B|^2)^n w = 0. \tag{5.4}$$

The equation

$$\left(\frac{\partial}{\partial \bar{z}} + q\frac{\partial}{\partial z}\right)^n w = 0, \tag{5.5}$$

is called the pluri-Beltrami equation.

First we consider the equation (5.1) for $n = 2$ and show that it is related to the equilibrium equations of elasticity theory written for stresses [Mu2]

$$\frac{\partial X_x}{\partial x} + \frac{\partial X_y}{\partial y} = 0,$$
$$\frac{\partial X_y}{\partial x} + \frac{\partial Y_y}{\partial y} = 0, \quad \Delta\Theta = 0, \quad \Theta = X_x + Y_y.$$

Because $\Theta - X_x = Y_y$, these equations have the form

$$\frac{\partial X_x - iX_y}{\partial \bar{z}} - \frac{1}{2}\frac{\partial \Theta}{\partial \bar{z}} + \frac{1}{2}\frac{\partial \Theta}{\partial z} = 0, \quad \text{i.e.,}$$
$$\frac{\partial X_x - Y_y - 2iX_y}{\partial \bar{z}} + \frac{\partial \Theta}{\partial z} = 0.$$

Thus we have

46   I. Two-Dimensional Cases

$$\frac{\partial^2(X_x - Y_y - 2iX_y)}{\partial \bar{z}^2} = 0.$$

From these equations follow

$$Y_y - X_x + 2iX_y = \bar{z}\Phi(z) + \psi(z),$$
$$X_x + Y_y = \varphi(z) + \overline{\varphi(z)},$$
$$\Phi(z) = \varphi'(z).$$

We obtain the Kolosov–Muskhelishvili formula

$$Y_y - X_x + 2iX_y = \bar{z}\varphi'(z) + \psi(z),$$
$$X_x + Y_y = \varphi(z) + \overline{\varphi(z)}. \tag{5.6}$$

From the equations of the plane theory of equilibrium of elastic bodies in the displacement components

$$(\lambda + \mu)\frac{\partial \theta}{\partial x_k} + \mu \Delta u_k = 0, \quad k = 1, 2, \quad \theta = \frac{\partial u_1}{\partial x_1} + \frac{\partial u_2}{\partial x_2},$$

one can get

$$(\lambda + \mu)\frac{\partial \theta}{\partial z} + 2\mu \frac{\partial^2 u_1 - iu_2}{\partial z \, \partial \bar{z}} = 0,$$
$$(\lambda + \mu)\theta + 2\mu \frac{\partial u_1 - iu_2}{\partial \bar{z}} = \overline{\psi'_1(z)},$$

where $\psi_1(z)$ is any holomorphic function of $z = x_1 + ix_2$. Using the last equation and its conjugate, one obtains

$$\theta = \frac{\partial u_1 - iu_2}{\partial \bar{z}} + \frac{\partial u_1 + iu_2}{\partial z} = \frac{1}{2(\lambda + 2\mu)}\left[\psi'_1(z) + \overline{\psi'_1(z)}\right].$$

By integrating with respect to $\bar{z}$, we obtain the well-known Kolosov–Muskhelishvili formula

$$2\mu(u_1 + iu_2) = \varkappa\varphi(z) - z\overline{\varphi'(z)} - \overline{\psi(z)},$$

where $\dfrac{\lambda + \mu}{2(\lambda + 2\mu)}\psi_1 \equiv \varphi(z)$, $\varkappa = \dfrac{\lambda + 3\mu}{\lambda + \mu} = 3 - 4\sigma$ is Muskhelishvili's constant.

## 5. BVP for Pluriholomorphic ... and the PluriBeltrami Equation

Now equations (5.5) with constant complex $q$, as in the case where $n = 1$, can be reduced to equation (5.1) by the transformation

$$\zeta = z - q\bar{z}. \tag{5.7}$$

Thus the solution of (5.5) can be represented as

$$w = \Phi(\zeta), \tag{5.8}$$

where $\Phi(\zeta)$ is a pluriholomorphic function.

First BVP are solved for pluriholomorphic functions. Let $D$ be the half plane $y > 0$.

**Problem 1.** Find the regular solution of (5.1) that vanishes at infinity by the conditions

$$\text{Re } \frac{\partial^k w}{\partial y^k}\bigg|_{y=0} = f_k(x), \quad k = 0, 1, \ldots, n-1. \tag{5.9}$$

**Solution.** The general solution of equation (5.1) can be written as

$$w(z) = \sum_{0}^{n-1} \bar{z}^k \varphi_k(z), \tag{5.10}$$

which can be also written in the form

$$w(z) = \sum_{0}^{n-1} (z - \bar{z})^k \psi_k(z) = \sum_{0}^{n-1} y^k \varphi_k(z), \tag{5.11}$$

where $\varphi_k(z)$ are any holomorphic functions.

Let $n = 2$. By conditions (5.9) we have

$$\text{Re } \varphi_0(x) = f_0(x), \quad \text{Re}\big[\varphi_1(x) + i\varphi_0'(x)\big] = f_1(x), \quad x \in R.$$

By these conditions the holomorphic functions $\varphi_0(z)$ and $\varphi_1(z) + i\varphi_0'(z)$ are defined as

$$\varphi_0(z) = \frac{1}{\pi i} \int_R \frac{f_0(t)dt}{t - z},$$

$$\varphi_1(z) + \frac{1}{\pi} \int_R \frac{f_0(t)dt}{(t-z)^2} = \frac{1}{\pi i} \int_R \frac{f_1(t)dt}{t-z}, \tag{5.12}$$

and $w(z)$ is defined by (5.11). The real part of the biharmonic function $u(x, y) = \operatorname{Re}[y\varphi_1(z) + \varphi_0(z)]$ is defined as

$$u(x, y) = \frac{2y^3}{\pi} \int_R \frac{f_0(t)dt}{[(x-t)^2 + y^2]^2} + \frac{y^2}{\pi} \int_R \frac{f_1(t)dt}{(t-x)^2 + y^2}, \qquad (5.13)$$

where $f_1$ must satisfy the condition

$$\int_R f_1(t)dt = 0 \qquad (5.14)$$

so that $u(x, y)$ vanishes as $y \to \infty$.

As was noted in the introduction, problems with the boundary condition $w(x, 0) = F(x)$ are not correctly posed. The representation (5.11) for $n = 2$ shows that, in this case, $\varphi_0(x, 0)$ is not right. A holomorphic function can be defined by the real part or the imaginary part given on the boundary [Mu1]. In problems with the boundary conditions

$$\operatorname{Re} w = f_0(x), \quad \operatorname{Im} \frac{\partial w}{\partial y} = f_1(x), \quad y = 0, \quad x \in R,$$

$\varphi_0(z)$ is represented as in (5.12) and $\varphi_1(z) + i\varphi'_0(z)$ as

$$\varphi_1(z) + i\varphi'_0(z) = \frac{1}{\pi} \int_R \frac{f_1(t)}{t - z} dt.$$

The biharmonic function $u(x, y)$ is defined by the given condition (5.9), $n = 2$, as

$$u(x, y) = \frac{2y^3}{\pi} \int_R \frac{f_0(t)dt}{[(t-x)^2 + y^2]^2} + \frac{y}{\pi} \int_R \frac{(t-x)f_1(t)dt}{(t-x)^2 + y^2}. \qquad (5.15)$$

By the conditions (5.9), in the case $n > 2$, one can define $\varphi_0(z)$, $\varphi_1 + i\varphi'_0(z)$, $2\varphi_2(z) + i\varphi'_1(z) - \varphi''_0(z)$, and so on as in (5.12). Then $u(x, y)$ is defined by (5.11), which can be written as

$$u(x, y) = \sum_{k=0}^{n-1} \frac{y^k}{k!} \sum_{m=0}^{k} (-1)^{k-m} \frac{d^{k-m}}{dy^{k-m}} \frac{y}{\pi} \int_R \frac{f_m(t)dt}{(x-t)^2 + y^2}. \qquad (5.16)$$

It is obvious that for this solution to vanish at infinity the given functions $f_k$, ($k = 1, \ldots, n-1$) must satisfy integral conditions like (5.14). Formula (5.16) was obtained in [Ob1] using FIT.

**Problem 2** (Riquie). Find the regular solution of (5.1) in $D$ that vanishes at infinity by the conditions

$$\operatorname{Re} \Delta^k w = f_k(x), \quad k = 0, 1, \ldots, n-1, \quad y = 0. \tag{5.17}$$

In the case of $n = 2$ by force of (5.11), (5.17) one has

$$\operatorname{Re} \varphi_0(x) = f_0, \quad \operatorname{Re}[2i\varphi_1'(x)] = f_1(x), \quad x \in R.$$

Thus $\varphi_0(z)$ and $2i\varphi_1'(z)$ are defined as in (5.12). One can obtain the representation for $w$. Then $u(x, y)$ is

$$u(x,y) = \frac{y}{\pi} \int_R \frac{f_0(t)dt}{(t-x)^2 + y^2} + \frac{y}{4\pi} \int_R f_1(t) \ln[(t-x)^2 + y^2] dt. \tag{5.18}$$

Similarly in the case of $n > 2$ the holomorphic functions $\varphi_k(z)$ $(k = 0, 1, \ldots, n-1)$ can be defined and $w$ has explicit representation. For its real part one obtains

$$u(x,y) = \frac{y}{\pi} \int_R \frac{f_0(t)dt}{(t-x)^2 + y^2} + \frac{y}{\pi} \sum_{k=1}^{n-1} \frac{2}{4^k[(k-1)!]^2 k} \int_R f_k(t) r^{2(k-1)} \ln r \, dt, \tag{5.19}$$

where $r^2 = (t - x)^2 + y^2$ and functions $f_k$ $(k = 1, \ldots, n-1)$ must satisfy the conditions

$$\int_R f_k(t) t^m dt = 0, \quad m = 0, 1, \ldots, 2(k-1).$$

The representation (5.19) was obtained in [Ob1] using FIT.

Now let $D$ be the circular domain $|z| \leq 1$ with the boundary $\Gamma : |z| = 1$.

**Problem 3.** Find the solution of (5.1) by the conditions

$$\operatorname{Re}\left[\frac{\partial^k w}{\partial r^k}\right]_{r=1} = f_k(t), \quad k = 0, 1, \ldots, n-1, \quad |t| = 1. \tag{5.20}$$

**Solution.** The solution of (5.1) can be represented as

$$w(z) = \sum_0^{n-1} (r^2 - 1)^k \varphi_k(z), \quad r = |z|. \tag{5.21}$$

50    I. Two-Dimensional Cases

First consider the case $n = 2$, i.e.,

$$w(z) = (r^2 - 1)\varphi_1(z) + \varphi_0(z). \tag{5.22}$$

By force of (5.20) for holomorphic functions $\varphi_0(z)$, $\varphi_1(z)$ one obtains the boundary conditions

$$\text{Re}[\varphi_0(t)] = f_0(t), \quad \text{Re}\left[2\varphi_1(t) + \frac{\partial \varphi_0}{\partial r}\right]_{r=1} = f_1(t). \tag{5.23}$$

Thus $\varphi_0(z)$ is defined by the Schwartz formula

$$\varphi_0(z) = \frac{1}{\pi i} \int_\Gamma \frac{f_0(t)dt}{t-z} - \frac{1}{2\pi i} \int_\Gamma f_0(t)\frac{dt}{t} + ic. \tag{5.24}$$

It is easy to show that $r\frac{\partial \varphi_0}{\partial r}$ is a holomorphic function which on $\Gamma$ is $\frac{\partial \varphi_0}{\partial r}$. Because

$$r\frac{\partial \varphi_0}{\partial r} = \frac{\partial \varphi_0}{\partial x}x + \frac{\partial \varphi_0}{\partial y}y = \left(\frac{\partial \varphi_0}{\partial \bar{z}} + \frac{\partial \varphi_0}{\partial z}\right)x - i\left(\frac{\partial \varphi_0}{\partial \bar{z}} - \frac{\partial \varphi_0}{\partial z}\right)y,$$

and $\frac{\partial \varphi_0}{\partial \bar{z}} = 0$, we have

$$r\frac{\partial \varphi_0}{\partial r} = z\varphi_0'(z). \tag{5.25}$$

That is why the holomorphic function $2\varphi_1(z) + z\varphi_0'(z)$ is also defined by (5.23) as the Schwartz formula

$$2\varphi_1(z) + z\varphi_0'(z) = \frac{1}{2\pi i}\int_\Gamma \frac{f_1(t)(t+z)}{t-z}\frac{dt}{t} + ic \equiv P(f_1).$$

Then by (5.22) one has the solution of (5.1)

$$w(z) = (r^2 - 1)P(f_1) + \varphi_0(z) - (r^2 - 1)z\varphi_0'(z). \tag{5.26}$$

By (5.24), the real part of this expression is a biharmonic function in $D$ which, by the corresponding boundary conditions, is defined as

$$u(x, y) = \frac{1}{2\pi}(r^2 - 1)^2 \left[ -\frac{1}{2}\int_0^{2\pi} \frac{f_1 d\vartheta}{1+r^2 - 2r\cos(\vartheta - \alpha)} \right.$$
$$\left. + \int_0^{2\pi} \frac{f_0(1 - r\cos(\alpha - \vartheta))d\vartheta}{[1+r^2 - 2r\cos(\vartheta - \alpha)]^2} \right].$$

## 5. BVP for Pluriholomorphic ... and the PluriBeltrami Equation

For $n > 2$ and by conditions (5.20) the holomorphic functions $\varphi_k(z)$ in (5.21) will be defined gradually.

First by the inductive method we can prove that if $\varphi(z)$ is a holomorphic function then for any $k$

$$r^k \frac{\partial^k \varphi}{\partial r^k} = z^k \varphi^{(k)}(z), \tag{5.27}$$

is holomorphic too. For $k = 1$ this was proved above. If it is true for $k - 1$, then it is true for $k$ since

$$r^k \frac{\partial^k \varphi}{\partial r^k} = r^k \frac{\partial}{\partial r} \frac{1}{r^{k-1}} r^{k-1} \frac{\partial^{k-1} \varphi}{\partial r^{k-1}} =$$

$$= r \frac{\partial}{\partial r} r^{k-1} \frac{\partial^{k-1} \varphi}{\partial r^{k-1}} - (k-1) r^{k-1} \frac{\partial^{k-1} \varphi}{\partial r^{k-1}} =$$

$$= z \frac{\partial}{\partial z} [z^{k-1} \varphi^{(k-1)}] - (k-1) z^{k-1} \varphi^{k-1}(z).$$

And we have the formula (5.27).

By force of (5.21) and for $r = 1$ one has the conditions

$$\text{Re}[\varphi_0(t)] = f_0(t), \quad \text{Re}\left[2\varphi_1(t) + \frac{\partial \varphi_0}{\partial r}\right] = f_1(t),$$

$$\text{Re}\left[8\varphi_2(t) + 2\varphi_1 + 2\frac{\partial \varphi_1}{\partial r} + \frac{\partial^2 \varphi_0}{\partial r^2}\right] = f_2(t),$$

and so on. By these conditions the holomorphic functions $\varphi_0(z)$, $2\varphi_1(z) + z\varphi_0'(z)$, $8\varphi_2(z) + 2\varphi_1(z) + 2z\varphi_1'(z) + z^2\varphi_0''(z)$, and so on will be defined as in (5.24). Thus the problem can be solved for any $n$.

**Problem 4** (Riquie). Find the regular solution of (5.1) in $D$ by the boundary conditions

$$\text{Re}[\Delta^k w]_{r=1} = f_k(t), \quad |t| = 1, \quad k = 0, 1, \ldots, n-1.$$

**Solution.** Again, representing $w(z)$ in the form (5.21), we can obtain corresponding conditions for the holomorphic functions $\varphi_k(z)$. For instance, in the case $n = 2$

$$\text{Re}[\varphi_0(t)] = f_0, \quad \text{Re}[(t\varphi_1)'] = \frac{1}{4} f_1 \equiv f, \quad |t| = 1.$$

Then $\varphi_0(z)$, $(z\varphi_1)'$ are defined by the Schwartz formula (5.24) and for $\varphi_1(z)$ one has

## I. Two-Dimensional Cases

$$\varphi_1(z) = \frac{1}{\pi i z} \int_\Gamma f(t) \ln(t-z) dt - \frac{1}{2\pi i} \int_\Gamma \frac{f(t)dt}{t} + ic.$$

It is clear that $f(t)$ must satisfy

$$\int_\Gamma f(t) \ln t \, dt = 0.$$

Thus one can see that problems for holomorphic functions in domains with cuts can be solved for pluriholomorphic functions (5.1) too using the representation (5.21).

By transformation (5.7) equation (5.5) with $q = const$ can be reduced to (5.1). Therefore, all problems solved for pluriholomorphic functions can be solved for pluri-Beltrami equation too.

Consider the bigeneralized holomorphic function $w(z)$

$$\frac{\partial w}{\partial \bar{z}} + B\bar{w} = F, \tag{5.28}$$

$$\frac{\partial F}{\partial \bar{z}} + B\bar{F} = 0, \tag{5.29}$$

where $B$ is real.

**Problem 5.** Find in the half plane $y > 0$ the regular solution $w(z)$ that vanishes at infinity by the conditions

$$\operatorname{Re} w(x,0) = \varphi(x), \quad \operatorname{Im}\left[\frac{\partial w}{\partial y}\right] = \psi(x), \quad y = 0. \tag{5.30}$$

**Solution.** By force of (5.28), (5.30) one can get

$$\operatorname{Re} F(x,0) = \varphi'(x) + B\varphi(x) - \psi(x) \equiv f(x).$$

Thus with this condition $F(x, y)$—the solution of (5.29)—is represented as in (2.22). Then $w$ is defined as the solution of the nonhomogeneous equation (5.28) with the condition $\operatorname{Re} w(x, 0) = \varphi(x)$ represented above.

Now consider the bimetaharmonic function.

**Problem 6.** Find in the half plane $y > 0$ the solution of the equation

$$(\Delta - k^2)^2 u = 0, \tag{5.31}$$

that vanishes at infinity by the conditions

## 5. BVP for Pluriholomorphic ... and the PluriBeltrami Equation

$$u(x,0) = f_1(x), \quad \frac{\partial^2 u}{\partial y^2} = f_2(x), \quad x \in R, \quad y = 0, \quad (5.32)$$

or

$$u(x,0) = f_1(x), \quad \frac{\partial u}{\partial y} = f_2(x), \quad x \in R, \quad y = 0. \quad (5.33)$$

**Solution.** Consider (5.31) with the conditions (5.32). These conditions are equivalent to the Riquie conditions

$$u(x,0) = f_1(x), \quad \Delta u = \varphi(x), \quad x \in R, \quad y = 0.$$

That means for the function $F(x, y)$

$$\begin{aligned}\Delta u - k^2 u &= F(x,y), \\ \Delta F - k^2 F &= 0,\end{aligned} \quad (5.34)$$

one has

$$F(x,0) = \varphi(x) - k^2 f_1(x).$$

Thus $F(x, y)$ is defined by force of (2.22) in quadratures. From (5.34) and by the condition $u(x, 0) = f_1(x)$ $u(x, y)$ is defined in quadratures too.

Now consider the boundary conditions (5.33). First it is easy to show that if $u_1, u_2$ are solutions of the equations

$$\Delta u_j - k^2 u_j = 0, \quad j = 1, 2, \quad (5.35)$$

then

$$u = y u_2 + u_1,$$

is a solution of equation (5.31). By force of (5.33) we obtain for $u_1, u_2$ the conditions

$$u_1(x,0) = f_1(x), \quad u_2(x,0) + \frac{\partial u_1}{\partial y} = f_2(x), \quad x \in R, \quad y = 0.$$

Thus, the solution $u_1, u_2$ of (5.35) is defined using (2.22).

All problems solved above for equation (5.35) can be solved for equation (5.31). In the same way for the equation

$$(\Delta - k^2)^n u(x, y) = 0,$$

with the conditions

$$\frac{\partial^p u}{\partial y^p} = f_p(x), \quad p = 0, 1, \ldots, n-1, \quad y = 0,$$

the solution can be represented in quadratures too.

## 6 Nonlocal Problems for Biholomorphic Functions

In [Ob1] certain nonlocal problems are solved for holomorphic and polyharmonic functions, for the plane theory of elasticity, and for other cases. It is possible to solve analogous problems for pluriholomorphic functions. Here we consider nonlocal problems for the biholomorphic equation

$$\frac{\partial^2 w}{\partial \bar{z}^2} = 0, \quad w = u + iv. \tag{6.1}$$

For higher order equations analogous problems can be solved in the same way.

Let $D$ be the half plane $y > 0$, $x \in R$.

**Problem 1.** Find the solution of (6.1) in $D$ that vanishes at infinity by the conditions

$$\frac{\partial u}{\partial y} = f_1(x), \quad y = 0, \quad x \in R, \tag{6.2}$$

and one of the nonlocal conditions:

$$\begin{aligned}(a) \quad u(x, 0) &= \lambda_1 u(x, h) + f_0(x), & x &> 0, \\ u(x, 0) &= f_0(x), & x &< 0, \end{aligned} \tag{6.3}$$

$$\begin{aligned}(b) \quad u(x, 0) &= \lambda_1 u(x, h_1) + f(x), & x &> 0, \\ u(x, 0) &= \lambda_2 u(x, h_2) + f(x), & x &< 0, \end{aligned} \tag{6.4}$$

where $h_1 \neq h_2, 0 < \lambda_1, \lambda_2 \leq 1$ are positive constants.

Problems with the conditions $\dfrac{\partial v}{\partial y} = f_1(x), y = 0, x \in R$ and (6.3) or (6.4) can be solved too.

## 6. Nonlocal Problems for Biholomorphic Functions

**Solution.** If we define $u(x, 0)$ for $x > 0$, the solution is represented by (5.13). Then for $u(x, 0) \equiv f_0(x)$, $x > 0$ by (6.3) one can obtain the Wiener–Hopf equation

$$f_0(x) + \frac{2h^3 \lambda_1}{\pi} \int_0^\infty \frac{f_0(t)\,dt}{[(x-t)^2 + h^2]^2} = g(x), \quad x > 0. \tag{6.5}$$

But in the case of (6.4) for the unknown function $u(x, 0) \equiv f_0(x)$, $x \in R$, one can use (5.13) to obtain the dual integral equations

$$f_0(x) + \frac{2h_1^3 \lambda_1}{\pi} \int_R \frac{f_0(t)\,dt}{[(x-t)^2 + h_1^2]^2} = g(x), \quad x > 0,$$

$$f_0(x) + \frac{2h_2^3 \lambda_2}{\pi} \int_R \frac{f_0(t)\,dt}{[(x-t)^2 + h_2^2]^2} = g(x), \quad x < 0. \tag{6.6}$$

To solve these equations we need the FIT of the kernel $k(x) = \frac{2\lambda_1 h^3}{\pi} \frac{1}{[x^2+h^2]^2}$, $x \in R$,

$$\widehat{k}(t) = \frac{2\lambda h^3}{\sqrt{2\pi}\,\pi} \int_R \frac{e^{ixt}\,dx}{[x^2 + h^2]^2} = -\frac{\lambda h^2}{\sqrt{2\pi}\,\pi} \frac{\partial}{\partial h} \int_R \frac{e^{ixt}\,dx}{x^2 + h^2}, \quad t \in R.$$

If we use Cauchy's integral formula one obtains

$$\widehat{k}(t) = \frac{\lambda h^2}{\sqrt{2\pi}} \frac{\partial}{\partial h} \left[ \frac{e^{-|t|h}}{h} \right], \quad h > 0. \tag{6.7}$$

Note that when FIT are used to solve integral equations of the convolution type it is convenient to consider the definition of FIT as

$$F_x[f(t)] \equiv \widehat{f}(x) = \frac{1}{\sqrt{2\pi}} \int_R f(t) e^{ixt}\,dt, \quad x \in R, \tag{6.8}$$

and the inverse as

$$f(x) = \frac{1}{\sqrt{2\pi}} \int_R \widehat{f}(t) e^{-ixt}\,dt, \quad x \in R. \tag{6.9}$$

Consider the right- and left-sided functions

$$f_+(x) = \begin{cases} f(x), & x > 0, \\ 0, & x < 0, \end{cases} \quad f_-(x) = \begin{cases} 0, & x > 0, \\ -f(x), & x < 0. \end{cases} \tag{6.10}$$

56   I. Two-Dimensional Cases

By (6.8) one can easily obtain

$$\widehat{f_+}(x) = \Phi^+(x), \quad \widehat{f_-}(x) = \Phi^-(x), \quad x \in R, \tag{6.11}$$

where $\Phi^+(x)$, $\Phi^-(x)$ are the boundary values of a function $\Phi(z)$ ($z = x + iy$) holomorphic in the upper half plane $y > 0$ and in the lower respectively. Namely, the definition of FIT is taken to be (6.8) to guarantee that the signs are in accordance in (6.11).

For the convolution of two functions

$$f * k = \frac{1}{\sqrt{2\pi}} \int_R f(t) k(x-t) dt, \tag{6.12}$$

we use the properties

$$F_x[f * k] = \widehat{f}(x) \cdot \widehat{k}(x),$$
$$F_x \Big[ \frac{1}{\sqrt{2\pi}} \int_R f(t) k(x+t) dt \Big] = \widehat{f}(-x) \widehat{k}(x). \tag{6.13}$$

Consider the Wiener–Hopf integral equation of convolution type with respect to $f(x)$,

$$f(x) + \frac{1}{\sqrt{2\pi}} \int_0^\infty k(x-t) f(t) dt = g(x), \quad x > 0, \tag{6.14}$$

where $k, g$ are given functions.

**Solution.** By introducing the one-sided functions $f_+(t) = f(t), t > 0$ and $f_-(t)$ to be defined below, equation (6.14) can be extended to negative $x$ too. In this way one gets the integral equation of convolution type defined on $R$

$$f_+(x) + \frac{1}{\sqrt{2\pi}} \int_R k(x-t) f_+(t) dt = g_+(x) + f_-(x), \quad x \in R, \tag{6.15}$$

where $f_-(x)$ is defined as

$$f_-(x) = \frac{1}{\sqrt{2\pi}} \int_R k(x-t) f_+(t) dt, \quad x < 0.$$

By FIT, and taking into consideration (6.11), we obtain the Hilbert BVP: find a piecewise holomorphic function $\Phi(z)$ that vanishes at infinity with jump line $R$ satisfying the boundary condition

$$\Phi^+(t) = G(t)\Phi^-(t) + g_1(t), \quad t \in R, \qquad (6.16)$$

where

$$G(t) = \left[1 + \sqrt{2\pi}\,\widehat{k}(t)\right]^{-1}, \quad g_1(t) = \left[1 + \sqrt{2\pi}\,\widehat{k}(t)\right]^{-1}\widehat{g}_+(t). \qquad (6.17)$$

In the case (6.7) it is easy to show that

$$1 + \sqrt{2\pi}\,\widehat{k}(t) = 1 - \lambda(1 + |t|h)e^{-h|t|} \geq 1 - \lambda, \quad \text{for } \lambda > 0.$$

Thus $[G(t)]^{-1} \neq 0$ for every $t \in R$, $0 < \lambda < 1$, and $\lim_{|t| \to \infty} G(t) = 1$. For this $\lambda$ the index of the problem (6.16) is

$$\varkappa = \frac{1}{2\pi}[\arg G(t)]_R = 0,$$

and (6.16) has a unique solution which is represented in quadratures [Mu1]. If $\Phi(z)$ is known the solution of equation (6.5) is defined

$$f_0(x) = f_+(x) = \frac{1}{\sqrt{2\pi}} \int_R \Phi^+(t)e^{-itx}dt, \quad x > 0.$$

This formula shows that the solution of (6.5) does not depend on $\Phi^-(t)$; i.e., it does not depend on the manner in which the equation is extended to negative $x$.

If $\lambda = 1$ then $G(0) = \infty$ and $G(t)$ has no other singularity, but the function

$$G_0(t) = \left(\frac{t}{t+i}\right)^2 G(t),$$

is bounded for every $t \in R$ and continuous at $t = 0$ too. Thus it is possible to consider (6.16) with $G_0(t)$ [Ga].

The solution of the dual integral equations (6.6) by FIT can be reduced to a BVP for holomorphic functions like (6.16). Thus it can be represented explicitly too.

Note that in [Ob1] there are nonlocal problems considered which are reduced to generalized Wiener–Hopf integral equations and to dual integral equations with two kernels depending on the difference and sum of the arguments

$$f(x) + \frac{1}{\sqrt{2\pi}} \int_0^\infty \left[k(x-t) + mk(x+t)\right]f(t)dt = g(x), \quad x > 0, \qquad (6.18)$$

58   I. Two-Dimensional Cases

and

$$f(x) + \frac{1}{\sqrt{2\pi}} \int_R \left[k_1(x-t) + mk_1(x+t)\right] f(t) dt = g_1(x), \quad x > 0,$$
$$f(x) + \frac{1}{\sqrt{2\pi}} \int_R \left[k_2(x-t) + mk_2(x+t)\right] f(t) dt = g_2(x), \quad x < 0,$$
(6.19)

where $m$ is a real number. The general form of (6.18), when the kernels have the form $K_1(x-t) + K_2(x+t)$, $K_1(x) \neq K_2(x)$, is investigated only in the sense of solvability [GC]. If $m = 0$ it is classical Wiener–Hopf equation and thus can be solved explicitly. We will see that in the cases $m = \pm 1$ it can also be solved explicitly. In applications we will have precisely these cases.

**Solution.** By introducing the one-sided functions $f_+(t) = f(t), t > 0$ and $f_-(t)$ defined analogously to (6.10), equation (6.18) can be extended to negative $x$ too. Thus, one gets the following integral equation defined on all $R$

$$f_+(x) + \frac{1}{\sqrt{2\pi}} \int_R [K(x-t) + mK(x+t)] f_+(t) dt = g_+(x) + f_-(x), \quad x \in R.$$
(6.20)

By force of (6.11), (6.13) and by applying FIT the solution of this equation is reduced to the BVP: find the piecewise holomorphic function $\Phi(z)$ that vanishes at infinity by the boundary condition

$$[1 + M(t)]\Phi^+(t) + mM(t)\Phi^+(-t) = \Phi^-(t) + \widehat{g}_+(t), \quad t \in R, \quad (6.21)$$

with $M(t) = \widehat{K}(t)$. Consider two piecewise holomorphic functions

$$\Phi_1(z) = \begin{cases} \Phi(z), & y > 0, \\ \Phi(-z), & y < 0, \end{cases} \quad \Phi_2(z) = \begin{cases} \Phi(-z), & y > 0, \\ \Phi(z), & y < 0. \end{cases} \quad (6.22)$$

From these equalities follow

$$\Phi_1(z) = \Phi_1(-z), \quad \Phi_2(z) = \Phi_2(-z). \quad (6.23)$$

Replacing $t$ by $-t$ in (6.21) and using (6.22), one can write these two conditions as a system of Hilbert BVP for $\Phi_1$ and $\Phi_2$

$$[1 + M(t)]\Phi_1^+(t) + mM(t)\Phi_1^-(t) = \Phi_2^-(t) + \widehat{g}_+(t),$$
$$[1 + M(-t)]\Phi_1^-(t) + mM(-t)\Phi_1^+(t) = \Phi_2^+(t) + \widehat{g}_+(-t), \quad t \in R.$$
(6.24)

## 6. Nonlocal Problems for Biholomorphic Functions 59

If the solutions of this system do not satisfy conditions (6.23), it is possible to construct other solutions for which they will be satisfied. It is easy to see that if $\Phi_1(z)$, $\Phi_2(z)$ are solutions of (6.24), then $\Psi_1(z) = \Phi_1(-z)$, $\Psi_2(z) = \Phi_2(-z)$ are solutions too. Using this property one can construct solutions with the help of their sum, which are even functions with respect to $z$, as required by (6.23).

As is known, the general solutions of Hilbert BVP for several piecewise holomorphic functions, i.e., for more than one boundary condition, cannot be represented in quadratures. Only in some partial cases when the solutions can be reduced to a Hilbert BVP for one certain function can they be represented explicitly. We will see that (6.24) can be solved in quadratures only if $m^2 = 1$. We will obtain precisely this case in the applications.

Let $1 + M(t) \neq 0$, $t \in R$. Rewrite (6.24) in the form

$$\Phi_1^+(t) = -\frac{mM(t)}{1+M(t)} \Phi_1^-(t) + \frac{1}{1+M(t)} \Phi_2^-(t) + g_1(t),$$
$$\Phi_2^+(t) = \frac{1+M(t)+M(-t)}{1+M(t)} \Phi_1^-(t) + \frac{mM(-t)}{1+M(t)} \Phi_2^-(t) + g_2(t), \quad t \in R. \qquad (6.25)$$

Then one can get ($t \in R$)

$$\Phi_1^+ + \Phi_2^+ = G(t)[\Phi_1^- + \Phi_2^-] + g_3(t) \quad \text{in the case } m = 1,$$
$$\Phi_1^+ - \Phi_2^+ = -G(t)[\Phi_1^- - \Phi_2^-] + g_4(t) \quad \text{in the case } m = -1, \qquad (6.26)$$

where

$$G(t) = \frac{1+M(-t)}{1+M(t)}. \qquad (6.27)$$

Hence $\Phi_1(z) + \Phi_2(z)$ or $\Phi_1(z) - \Phi_2(z)$ is defined as the solution of the Hilbert BVP (6.26) for one piecewise holomorphic function; i.e., it is represented in quadratures. Then define, for instance, $\Phi_2(z)$ by means of $\Phi_1(z)$ and substitute it into the first equation of (6.25). In this way one gets very simple conditions to define $\Phi_1(z)$

$$\Phi_1^+ + \Phi_1^- = g_5(t) \quad \text{in the case } m = 1,$$
$$\Phi_1^+ - \Phi_1^- = g_6(t) \quad \text{in the case } m = -1, \quad t \in R. \qquad (6.28)$$

Thus $\Phi_1(z)$ and $\Phi_2(z)$ can be represented immediately by Cauchy-type integrals. The functions $g_k(t)$, $k = 1, \ldots, 6$, in the above equalities can be easily defined by $g(x)$ given in (6.18). Thus, $\Phi(z)$ is defined by (6.22) and the solution $f(x)$ of (6.18) by force of (6.11) is represented as (6.8).

It is remarkable that in applications $K(x)$ is an even function. Thus by force of (6.27) we have $G(t) = 1$ and problems with the boundary conditions (6.26) are, like (6.28), very simple.

60    I. Two-Dimensional Cases

Now consider the dual integral equations (6.19). The case of $m = 0$ has already been considered above as (6.6). Thus consider $m \neq 0$. If $K_1(x) = K_2(x)$, $x \in R$, equations (6.19) are the simplest. In fact we have one equation for $x \in R$, and using the FIT, we can obtain a linear equation with respect to $\widehat{f}(t)$ and $\widehat{f}(-t)$. Moreover, the second equation is obtained from this equation by changing $t$ to $-t$. Supposing that its determinant

$$1 + M(t) + M(-t) + (1 - m^2)M(t)M(-t) \neq 0, \quad M(t) = \widehat{K}_1(t), \quad t \in R,$$

$\widehat{f}(t)$ is defined uniquely, then by the inverse formula, $f(t)$ is defined too. Thus later on we consider

$$m \neq 0, \quad K_1(x) \neq K_2(x). \tag{6.29}$$

By the additional right- and left-sided functions equations (6.19) can be written as

$$f(x) + \frac{1}{\sqrt{2\pi}} \int_R f(t)[K_1(x-t) + mK_1(x+t)]dt = g_1(x) + f_-(x),$$

$$f(x) + \frac{1}{\sqrt{2\pi}} \int_R f(t)[K_2(x-t) + mK_2(x+t)]dt = g_2(x) + f_+(x), \quad x \in R.$$

Applying FIT to this system and taking into consideration (6.11), (6.13) we obtain the system of equations

$$\begin{aligned} [1 + M_1(t)]\widehat{f}(x) + mM_1(x)\widehat{f}(-x) = \widehat{g}_1(x) + \Phi^-(x), \\ [1 + M_2(t)]\widehat{f}(x) + mM_2(x)\widehat{f}(-x) = \widehat{g}_2(x) + \Phi^+(x), \quad x \in R, \end{aligned} \tag{6.30}$$

where $M_1 = \widehat{K}_1$, $M_2 = \widehat{K}_2$. By force of conditions (6.29)

$$D(x) \equiv m[M_2(x) - M_1(x)] \neq 0. \tag{6.31}$$

Therefore, from the last equations one can define $\widehat{f}(x)$ and $\widehat{f}(-x)$ uniquely. Replace $-x$ by $x$ in the boundary conditions for the piecewise holomorphic function $\Phi(z)$ that vanishes at infinity to get

$$mD(-x)[M_2(x)\Phi^-(x) - M_1(x)\Phi^+(x)] =$$
$$= D(x)[(1 + M_1(-x))\Phi^+(-x) - (1 + M_2(-x))\Phi^-(-x)] + D(x)D(-x)g(x), \tag{6.32}$$

where $g(x)$, $x \in R$, is defined easily by $\widehat{g}_1$ and $\widehat{g}_2$. Consider the two piecewise holomorphic functions $\Phi_1(z)$, $\Phi_2(z)$ defined by (6.28). Then (6.30) and the equality obtained from (6.20), where $x$ is replaced by $-x$, can be rewritten as

$$mM_1(x)D(-x)\Phi_1^+(x) - [1 + M_2(-x)]D(x)\Phi_2^+(x) =$$
$$= -[1 + M_1(-x)]D(x)\Phi_1^-(x) + mM_2(x)D(-x)\Phi_2^-(x) - D(x)D(-x)g(x),$$
$$[1 + M_1(x)]D(-x)\Phi_1^+(x) - mM_2(-x)D(x)\Phi_2^+(x) = \qquad (6.33)$$
$$= -mM_1(-x)D(x)\Phi_1^-(x) + [1 + M_2(x)]D(-x)\Phi_2^-(x) -$$
$$- D(x)D(-x)g(-x), \quad x \in R.$$

Thus we have the Hilbert problem for two piecewise holomorphic functions $\Phi_1(z)$, $\Phi_2(z)$ that vanishes at infinity. In general this system cannot be solved explicitly. But when $m^2 = 1$ it can. We will consider just this case. Let

$$D_1(x) = 1 + M_1(x) + M_2(-x) \neq 0, \quad x \in R. \qquad (6.34)$$

Then the system (6.33) can be written in an equivalent form

$$\Phi_1^+(x) = \frac{1}{D_1(x)}\left[D(x)\Phi_1^- + (1 + M_2(x) + M_2(-x))\Phi_2^-\right] + d_1(x),$$
$$\Phi_2^+(x) = \frac{1}{D_1(x)}\left[(1 + M_1(x) + M_1(-x))\Phi_1^- - D(-x)\Phi_2^-\right] + d_2(x). \qquad (6.35)$$

It is easy to obtain a Hilbert BVP for one piecewise holomorphic function by taking into consideration (6.31)

$$\Phi_1^+(x) + \Phi_2^+(x) = G(x)[\Phi_1^-(x) + \Phi_2^-(x)] + d(x), \quad x \in R. \qquad (6.36)$$

In the cases $m = 1$ and $m = -1$ we have

$$\Phi_1^+(x) - \Phi_2^+(x) = -G(x)[\Phi_1^-(x) - \Phi_2^-(x)] + d(x), \quad x \in R, \qquad (6.37)$$

where

$$G(x) = \frac{1 + M_2(x) + M_1(-x)}{1 + M_2(-x) + M_1(x)}. \qquad (6.38)$$

By force of (6.34) $G(x) \neq 0$, $x \in R$, and $\lim_{|x|\to\infty} G(x) = 1$. Thus by conditions (6.36) and (6.37) the holomorphic functions $\Phi_1(z) + \Phi_2(z)$ in the case $m = 1$ and

$\Phi_1(z) - \Phi_2(z)$ in the case $m = -1$ are defined in quadratures. Then, defining $\Phi_2(z)$ with the help of $\Phi_1(z)$ and substituting it into the first condition (6.35), we obtain the Hilbert BVP again for one piecewise holomorphic function $\Phi_1(z)$ that vanishes at infinity with the conditions (6.28). Therefore $\Phi_1(z)$ and $\Phi_2(z)$ are defined perfectly. Next $\Phi(z)$ is defined by (6.22) and $\widehat{f}(x)$ is defined by (6.30). Hence, the solution of the dual integral equations (6.19) is represented by $\widehat{f}(x)$ using the inverse formula.

It is remarkable that in some applications $K_1(x)$ and $K_2(x)$ are even functions. Thus, by force of (6.38) one has $G(x) = 1$ and the solution of (6.36) or (6.37) that vanishes at infinity has a very simple representation as a Cauchy-type integral.

Let $D$ be the half plane $y > 0$, $x \in R$, and $D_+$ be the quarter of the plane $x > 0$, $y > 0$. Using the properties of Cauchy-type integrals, the holomorphic function in $D$ function $\Psi(z) = u(x, y) + iv(x, y)$ that vanishes at infinity by the condition

$$u(x, 0) = f(x), \quad x \in R, \tag{6.39}$$

is represented as

$$\Psi(z) = \frac{1}{\pi i} \int_R \frac{f(t)dt}{t - z}, \quad z = x + iy, \quad y > 0, \tag{6.40}$$

and the holomorphic function in $D_+$ $\psi(z)$ that vanishes at infinity by the conditions

$$u(x, 0) = f(x), \quad x > 0, \quad u(0, y) = g(y), \quad y > 0, \tag{6.41}$$

or

$$u(x, 0) = f(x), \quad x > 0, \quad v(0, y) = g(y), \quad y > 0, \tag{6.42}$$

can be represented, using the symmetric principle of reflection, as

$$\Psi(z) = \frac{1}{\pi i} \int_0^\infty \left[ \frac{1}{t - z} \pm \frac{1}{t + z} \right] f(t)dt + \frac{1}{\pi k} \int_0^\infty \left[ \frac{1}{t - z} \pm \frac{1}{t + z} \right] g(t)dt, \tag{6.43}$$

where in the case (6.41), the sign $(+)$ and $k = i$ must be taken, whereas in the case (6.42), the sign $(-)$ and $k = 1$ must be taken.

It is easy to define the FIT of the kernel

$$K(x) = \sqrt{\frac{2}{\pi}} \frac{h}{h^2 + x^2}, \quad \widehat{K}(t) = e^{-h|t|}, \quad h > 0. \tag{6.44}$$

**Problem.** Find the holomorphic function in $D_+$ $\Psi(z)$ that vanishes at infinity by the conditions

## 6. Nonlocal Problems for Biholomorphic Functions

$$u(x, 0) = \lambda u(x, h) + \varphi(x), \quad x > 0, \quad u(0, y) = g(y), \quad y > 0, \quad (6.45)$$

or

$$u(x, 0) = \lambda u(x, h) + \varphi(x), \quad x > 0, \quad v(0, y) = g(y), \quad y > 0. \quad (6.46)$$

**Solution.** These conditions combined with (6.43) give for the unknown function $u(x, 0) = f(x)$, $x > 0$, the generalized Wiener–Hopf equation (6.18)

$$f(x) - \frac{1}{\sqrt{2\pi}} \int_0^\infty [K(x-t) \mp K(x+t)] f(t) dt = \varphi_0(t), \quad x > 0, \quad (6.47)$$

where the sign ($-$) applies in the case (6.45) and the sign ($+$) applies in the case (6.46) and the kernel $K(x)$ is defined by (6.44). By force of (6.26), (6.27) these equations reduce to the boundary problem for holomorphic functions with the condition

$$\Phi_1^+ \pm \Phi_2^+ = \pm G(t)[\Phi_1^- \pm \Phi_2^-] + g_1(t), \quad t \in R, \quad (6.48)$$

where by force of (6.44) $G(t) = 1$; i.e., $\Phi_1(z) \pm \Phi_2(z)$ is defined uniquely by a Cauchy-type integral. In light of conditions (6.25), $\Phi_1(z)$ has simple conditions like (6.48) with $G(t) = 1$. Thus it is also defined by a Cauchy-type integral. Then $f(x)$ is represented by the inverse formula.

Note that the problem for $\Psi(z)$ in $D$ with the conditions

$$u(x, 0) = \varphi(x), \quad x < 0, \quad u(x, 0) = u(x, h) + mu(-x, h) + g(x), \quad x > 0, \quad (6.49)$$

where $h > 0$, $m^2 = 1$, can be solved explicitly with the help of (6.47).

**Problem.** Find the holomorphic function in $D$ $\Psi(z)$ that vanishes at infinity by the conditions

$$\begin{aligned} u(x, 0) &= u(x, h_1) + mu(-x, h_1) + \varphi_1(x), \quad x > 0, \\ u(x, 0) &= u(x, h_2) + mu(-x, h_2) + \varphi_2(x), \quad x < 0, \end{aligned} \quad (6.50)$$

where $m^2 = 1$, $h_1 \neq h_2$ are positive numbers. The case $h_1 = h_2$ is simple and the solution can be written easily.

**Solution.** By virtue of (6.40), (6.50) one can obtain for the unknown function $u(x, 0) \equiv f(x)$, $x \in R$, dual integral equations of the form (6.19), where $K_1(x)$ and $K_2(x)$ are defined by (6.44). Because the FIT of these kernels, by force of (6.44), are even functions, the BVP (6.36), (6.37) have the simplest form with $G(x) = 1$. Thus $\Phi_1(z)$, $\Phi_2(z)$ are defined by Cauchy-type integrals. Then $f(x)$ is defined uniquely too.

Analogous problems can be formulated for the biholomorphic function (6.1) which are solved using the integral equations (6.18), (6.19).

## 7 More BVP for Pluriholomorphic and Plurigeneralized Holomorphic Functions

Let $\Gamma$ be a set of finitely many smooth simple closed curves in the complex plane $z = x + iy$ and $w(z)$ be a piecewise pluriholomorphic function with jump line $\Gamma$.

**Hilbert Problem.** Find the solution of the equation

$$\frac{\partial^n w}{\partial \bar{z}^n} = 0, \quad n \geq 1, \tag{7.1}$$

that vanishes at infinity and satisfies the boundary conditions

$$\left(\frac{\partial^k w}{\partial \bar{z}^k}\right)^+ - G_k \left(\frac{\partial^k w}{\partial \bar{z}^k}\right)^- = g_k(t), \quad t \in \Gamma, \quad k = 0, 1, \ldots, n-1, \tag{7.2}$$

where $G_k(t)$, $g_k$ are given Hölder-continuous functions.

**Solution.** In the case $n = 1$ we have the classical Hilbert problem for holomorphic functions, the solution of which is well known.

Consider $n = 2$. By the condition for $\frac{\partial w}{\partial \bar{z}}$, that case can be defined as a piecewise holomorphic function, i.e.,

$$\frac{\partial w}{\partial \bar{z}} = \phi(z). \tag{7.3}$$

Then one can obtain

$$w = \bar{z}\Phi(z) + \Psi(z), \tag{7.4}$$

where the piecewise holomorphic function $\Psi(z)$ is defined as the solution of the Hilbert problem

$$\Psi^+(t) - G_0 \Psi^-(t) = g_0(t) - \bar{t}(\Phi^+ - G_0 \Phi^-) = g(t). \tag{7.5}$$

For $n > 2$ the solution of (7.1) is represented as

$$w(z) = \sum_{0}^{n-1} \bar{z}^k \Phi_k(z), \tag{7.6}$$

where $\Phi_k(z)$ are piecewise holomorphic functions that are defined gradually by conditions (7.2).

## 7. More BVP for Pluriholomorphic and Plurigeneralized Holomorphic Functions

**Problem 2** (Compound BVP). Let $D$ be a domain bounded by a closed smooth line $L$ inside of which lines $\Gamma$ are situated. Find the piecewise pluriholomorphic function in $D$ $w(z)$ with jump line $\Gamma$ by the conditions

$$\left(\frac{\partial^k w}{\partial \bar{z}^k}\right)^+ - G_k \left(\frac{\partial^k w}{\partial \bar{z}^k}\right)^- = g_k(t), \quad k = 0, \ldots, n-1, \quad t \in \Gamma, \tag{7.7}$$

$$\operatorname{Re}\left[\lambda_k(t) \frac{\partial^k w}{\partial \bar{z}^k}\right]_{z \to t} = \varphi_k(t), \quad k = 0, \ldots, n-1, \quad t \in L. \tag{7.8}$$

**Solution.** To solve this problem in quadratures suppose that $L$ is the circle $|z| = 1$. Consider $n = 2$. Then by condition (7.7) for $\frac{\partial w}{\partial \bar{z}}$ that case can be defined as a piecewise holomorphic function in $D$ with jump line $\Gamma$ in the form

$$\frac{\partial w}{\partial \bar{z}} = \Phi_1(z) + \Phi_2(z), \tag{7.9}$$

where $\Phi_1(z)$ is the piecewise holomorphic function in $D$ with jump line $\Gamma$ defined by the condition

$$\Phi_1^+ - G_1 \Phi_1^- = g_1,$$

and $\Phi_2(z)$ is any holomorphic function in $D$. Thus $\Phi_1(z)$ is defined explicitly and $\Phi_2(z)$ is defined using (7.9) and (7.8) for $\frac{\partial w}{\partial \bar{z}}$. In this way we obtain for $\Phi_2(z)$ the Riemann–Hilbert boundary condition

$$\operatorname{Re}[\lambda_1(t) \Phi_2(t)] = \varphi(t), \tag{7.10}$$

and it is represented in quadratures. Then from (7.9) it follows that

$$w(z) = \bar{z}(\Phi_1(z) + \Phi_2(z)) + \Psi(z), \tag{7.11}$$

where $\Psi(z)$ is a piecewise holomorphic function with jump line $\Gamma$. It is defined analogously by the conditions (7.7), (7.8) for $k = 0$. For $n > 2$ the solution of (7.7), (7.8) is defined gradually.

Note that the Liouville theorem which is well known for holomorphic functions can be formulated for pluriholomorphic functions in the following form.

**Liouville theorem.** *Let $w(z)$ be the solution of equation (7.1) in the complex plane that satisfies the conditions*

$$\lim_{|z| \to \infty} \frac{\partial^k w}{\partial \bar{z}^k} = 0, \quad k = 0, 1, \ldots, n-1. \tag{7.12}$$

*Then $w(z) = 0$ everywhere.*

The proof is obvious.

Now let $w(z)$ be a plurigeneralized holomorphic function, i.e., a solution of the equations

$$P^n w = 0, \quad Pw = \frac{\partial w}{\partial \bar{z}} + B\bar{w}. \tag{7.13}$$

The above two problems can be considered for this equation too. But to solve it explicitly, $D$ will be taken to be the half plane $y > 0$ and $B$ is a complex constant.

**Hilbert Problem.** Find the piecewise plurigeneralized holomorphic function with jump line $\Gamma$ that vanishes at infinity by the conditions

$$\left(\frac{\partial^k w}{\partial \bar{z}^k}\right)^+ - \left(\frac{\partial^k w}{\partial \bar{z}^k}\right)^- = g_k(t), \quad t \in \Gamma, \quad k = 0, \ldots, n-1. \tag{7.14}$$

**Compound BVP.** Find the solution of equation (7.13) in $D$ with jump line $L$ that vanishes at infinity by the conditions

$$\left(\frac{\partial^k w}{\partial \bar{z}^k}\right)^+ - \left(\frac{\partial^k w}{\partial \bar{z}^k}\right)^- = g_k(t), \quad t \in \Gamma, \quad k = 0, \ldots, n-1,$$

$$\operatorname{Re}\left[\frac{\partial^k w}{\partial \bar{z}^k}\right]_{z \to t} = \varphi_k(t), \quad t \in L. \tag{7.15}$$

Because we know the solution of these problems for generalized holomorphic functions in quadratures, it is obvious that in this case for pluriholomorphic functions the solution is represented explicitly too.

# II
# Multidimensional Cases

## 0  Introduction

The equations considered in the following can be obtained by applying the Dirac operator in Clifford analysis. They are related to polyharmonic, polymetaharmonic, and poly-Beltrami equations in multidimensional spaces, all of which have remarkable applications. To solve for them boundary value problems (BVP) in quadratures and sometimes Fourier integral transformations (FIT) will be used. First some basic notions and definitions will be described.

**0.1  Elements of Clifford analysis (e.g., [Ob1]).** Let $R_{(n)}$, $R_{(n,n-1)}$ and $R_{(n)}^0$ ($n \geq 1$) be Clifford algebras with the basis $\{e_A\}$, $A = (\alpha_1, \ldots, \alpha_k)$ with $1 \leq \alpha_1 < \cdots < \alpha_k \leq n$, and with the multiplication rules

$$\begin{cases} e_0^2 = e_0, \quad e_j^2 = -e_0 \text{ for } j = 1, \ldots, n-1, \\ e_j e_k + e_k e_j = 0 \text{ for } j, k = 1, \ldots, n \text{ and } j \neq k, \end{cases} \quad (1)$$

$$e_n^2 = -e_0 \text{ in the case } R_{(n)}, \quad (2)$$

$$e_n^2 = e_0 \text{ in the case } R_{(n,n-1)}, \quad (3)$$

$$e_n^2 = 0 \text{ in the case } R_{(n)}^0,$$

where $e_0$ is the identity element. Thus, these spaces are associative, $2^n$-dimensional as real spaces, and noncommutative (for $n \geq 2$). Any element can be represented as

$$u = \sum_A u_A e_A, \qquad (4)$$

where $e_A = e_{\alpha_1} \cdots e_{\alpha_k}$. An element $u$ is vectorial if

$$u = \sum_{k=0}^n u_k e_k. \qquad (5)$$

For every $u$ two conjugates are defined:

$$\bar{u} = \sum_A u_A \bar{e}_A, \quad \tilde{u} = \sum_A u_A \tilde{e}_A, \qquad (6)$$

where $\bar{e}_0 = e_0, \bar{e}_j = \tilde{e}_j = -e_j, j = 1, \ldots, n$, and

$$\bar{e}_A = \bar{e}_{\alpha_k} \cdots \bar{e}_{\alpha_1} = (-1)^{k(k+1)/2} e_A, \quad \tilde{e}_A = \tilde{e}_{\alpha_1} \cdots \tilde{e}_{\alpha_k} = (-1)^k e_A. \qquad (7)$$

$R_{(1)}$ is the space of complex numbers, $R_{(1,0)}$ is the space of double numbers, and $R_{(1)}^0$ is the space of dual numbers.

Consider the modification of the Dirac operator [BDS],

$$\bar{\partial} = \sum_{k=0}^n \frac{\partial}{\partial x_k} e_k = \frac{\partial}{\partial x_0} e_0 + D, \quad \partial = \frac{\partial}{\partial x_0} e_0 - D, \quad x = \sum_{k=0}^n x_k e_k, \qquad (8)$$

where $D$ is the Dirac operator. One has in $R_{(n)}$, $R_{(n,n-1)}$ and $R_{(n)}^0$, correspondingly,

$$\partial \bar{\partial} = \bar{\partial} \partial = \Delta_{(n)}, \quad \partial \bar{\partial} = \Delta_{(n-1)} - \frac{\partial^2}{\partial x_n^2}, \quad \partial \bar{\partial} = \Delta_{(n-1)}, \qquad (9)$$

where $\Delta_{(k)}$ ($k = n, n-1$) is the Laplace operator with respect to the variables $x_0, \ldots, x_k$. From these equalities it follows that the equation

$$\bar{\partial} u + \tilde{u} h = 0, \qquad (10)$$

is elliptic in $R_{(n)}$ and hyperbolic in $R_{(n,n-1)}$. In these spaces the Beltrami and generalized Beltrami equations

$$\bar{\partial} u + q \partial u = 0, \qquad (11)$$
$$\bar{\partial} u + q_1 \partial u + q_2 \bar{\partial} \bar{u} = 0, \qquad (12)$$

will be considered. We will see that these equations can be elliptic or hyperbolic depending on the conditions imposed on the coefficients.

Besides the pluriregular and pluri-Beltrami equations in $R_{(n)}$ and $R_{(n,n-1)}$ we will consider

$$\bar{\partial}^m u = 0, \tag{13}$$
$$(\bar{\partial} + q\partial)^m u = 0. \tag{14}$$

**0.2 The basic $L$ theory of the FIT.** In the theory of PDEs the FIT is widely used as an effective method of constructing the solutions of boundary and initial value problems (B&IVP) in quadratures. The FIT is defined on various spaces of functions. Since our purpose is to represent the solution explicitly, it is sufficient to consider the FIT on the space $L$, i.e., the space of all measurable functions $f(x)$, in general complex valued and defined on the $n$-dimensional Euclidean space $R^n$ ($n \geq 1$) with the $L$-norm

$$\|f(x)\|_L = \int_{R^n} |f(x)|\, dx < \infty. \tag{15}$$

The simplest properties of the FIT will be considered below.

The FIT of $f(x) \in L(R^n)$ is the function $\widehat{f}(y)$ defined by

$$F_y[f(x)] \equiv \widehat{f}(y) = \frac{1}{(\sqrt{2\pi})^n} \int_{R^n} f(x) e^{-i(x \cdot y)}\, dx \quad \text{for all } y \in R^n, \tag{16}$$

where $x \cdot y = \sum_1^n x_j y_j$, $dx = dx_1\, dx_2 \cdots dx_n$.

Taking into consideration Fubini's theorem about the inversion of the order of integration, $\widehat{f}(y)$ can be represented as repeated integrals

$$\widehat{f}(y) = \frac{1}{(\sqrt{2\pi})^n} \int_R e^{-ix_1 y_1}\, dx_1 \cdots \int_R f(x) e^{-ix_n y_n}\, dx_n. \tag{17}$$

Then if $f(x) = \prod_1^n f_k(x_k)$, one has

$$\widehat{f}(y) = \prod_1^n \widehat{f_k}(y_k). \tag{18}$$

Let $f(x), g(x) \in L(R^n)$. Then $\widehat{f}g$ and $f\widehat{g}$ are in the same class $L(R^n)$ and

$$\int_{R^n} \widehat{f}(x) g(x) e^{i(x \cdot y)}\, dx = \int_{R^n} f(x) \widehat{g}(x-y)\, dx. \tag{19}$$

Its proof is very simple using Fubini's theorem. The last equality with $y = 0$ is the Parseval formula.

Let $H(x)$, $K(x) = \widehat{H} \in L(R^n)$, where $H(x)$ is continuous in the neighborhood of $x = 0$, $H(0) = 1$, $K(-x) = K(x)$, and

$$\int_{R^n} K(x)\,dx = (\sqrt{2\pi})^n. \tag{20}$$

Then $H$ and $K$ are called a pair of Fejér kernels. It is easy to show that

$$F_y\left[H\left(\frac{x}{\rho}\right)\right] = \rho^n K(\rho y). \tag{21}$$

**Inversion theorem.** Let $f, \widehat{f} \in L(R^n)$. Then almost everywhere in $R^n$

$$f(x) = \frac{1}{(\sqrt{2\pi})^n} \int_{R^n} \widehat{f}(y) e^{i(x \cdot y)}\,dy, \tag{22}$$

and, moreover, if $f \in C(R^n)$, this equality is true everywhere.

Consider (19) for any $f(x) \in L$ and $g(x) = H(x/\rho)$ defined above. Then by (21) one can get

$$\int_{R^n} \widehat{f}(y) H\left(\frac{y}{\rho}\right) e^{i(x \cdot y)}\,dy = \int_{R^n} f(y) \rho^n K(\rho(y - x))\,dy =$$

$$= \int_{R^n} f\left(x + \frac{y}{\rho}\right) K(y)\,dy.$$

Let $f(x) \in C(R^n) \cap L(R^n)$. Then using Lebesgue's convergence theorem and (20)

$$\lim_{\rho \to \infty} \int_{R^n} f\left(x + \frac{y}{\rho}\right) K(y)\,dy = (\sqrt{2\pi})^n f(x).$$

Thus equality (22) is valid everywhere.

Let $f(x), g(x) \in L(R^n)$. Then the usual product $f(x)g(x)$ is not generally in the space $L(R^n)$. The 'product' in this space is defined in such a way that $L(R^n)$ is a Banach algebra. This operator, called convolution, is defined as

$$h(x) = \frac{1}{(\sqrt{2\pi})^n} \int_{R^n} f(x - y) g(y)\,dy = f * g, \quad x \in R^n. \tag{23}$$

By Fubini's theorem we have $h(x) \in L(R^n)$ and

$$\|h(x)\|_L \leq \|f\|_L \|g\|_L.$$

Moreover, this operator is commutative and associative, and

$$\widehat{f*g} = \hat{f}\cdot\hat{g}. \tag{24}$$

Let $f, g, \hat{f} \in L(R^n)$. Then $f*g$ and $\hat{f}\cdot\hat{g}$ are functions of $L(R^n)$, so that by the inversion theorem (22) and by (24), it follows that

$$\frac{1}{(\sqrt{2\pi})^n}\int_{R^n}\hat{f}\hat{g}e^{-i(x\cdot y)}dx = \frac{1}{(\sqrt{2\pi})^n}\int_{R^n}f(x)g(y-x)\,dx. \tag{25}$$

Let $f(x) \in L(R^n)$ $(n \geq 1)$ depend only on $|x| \equiv \tau$. Then $\hat{f}(y)$ is dependent only on $|y| \equiv \rho$. For $n=1$ this is obvious because in this case

$$\hat{f}(y) = \frac{1}{\sqrt{2\pi}}\int_R f(|x|)\cos xy\,dx.$$

For $n \geq 2$ consider spherical coordinates in $R^n$ (polar coordinates for $n=2$):

$$\begin{aligned}
x_1 &= r\sin\phi_1\sin\phi_2\cdots\sin\phi_{n-1},\\
x_2 &= r\sin\phi_1\sin\phi_2\cdots\sin\phi_{n-2}\cos\phi_{n-1},\\
x_3 &= r\sin\phi_1\sin\phi_2\cdots\sin\phi_{n-3}\cos\phi_{n-2},\\
&\cdots\cdots\cdots\cdots\cdots\cdots\cdots\cdots\cdots\cdots\\
x_{n-1} &= r\sin\phi_1\cos\phi_2,\\
x_n &= r\cos\phi_1,
\end{aligned} \tag{26}$$

where $0 \leq \phi_{n-1} \leq 2\pi$, $0 \leq \phi_k \leq \pi$, $k=1,\ldots,n-2$. Then

$$dx = r^{n-1}\sin^{n-2}\phi_1\sin^{n-3}\phi_2\cdots\sin\phi_{n-2}\,dr\,d\phi_1\cdots d\phi_{n-1} =$$
$$= r^{n-1}\sin^{n-2}\phi_1\,dr\,d\phi_1\,d\omega_{n-1},$$

where $d\omega_{n-1}$ is a volume element of the unit sphere in $R^{n-1}$. As the axis $x_n$ can be taken along $y$, $x\cdot y = r\rho\cos\phi_1$. Therefore (16) is represented in the form

$$\hat{f}(y) = \frac{\omega_{n-1}}{(\sqrt{2\pi})^n}\int_0^\infty f(r)r^{n-1}dr\int_0^\pi e^{-ir\rho\cos\phi}\sin^{n-2}\phi\,d\phi, \tag{27}$$

where $\omega_n$ is the surface area of the unit sphere in $R^n$ and

$$\omega_n = 2\pi^{n/2}\left[\Gamma\left(\frac{n}{2}\right)\right]^{-1},\quad \Gamma\left(\frac{1}{2}\right) = \sqrt{\pi}, \tag{28}$$

## II. Multidimensional Cases

where $\Gamma(n) = (n-1)\Gamma(n-1)$ is the Euler function of the second kind. This equality demonstrates that the theorem is true. Now we obtain the formula in order to calculate it.

It is known that

$$\frac{1}{\pi}\int_0^\pi e^{-ir\rho\cos\phi}\,d\phi = J_0(r\rho), \tag{29}$$

where $J_0$ is the Bessel function of zeroth order. Apart from this, one can easily obtain

$$\int_0^\pi e^{-ir\rho\cos\phi}\sin\phi\,d\phi = \frac{2\sin r\rho}{r\rho}. \tag{30}$$

Hence,

$$\widehat{f}(\rho) = \int_0^\infty f(r)r J_0(r\rho)\,dr \quad \text{for} \quad n=2,$$
$$\widehat{f}(\rho) = \frac{2}{\sqrt{2\pi}\,\rho}\int_0^\infty f(r)r\sin r\rho\,dr \quad \text{for} \quad n=3. \tag{31}$$

For $n > 3$ consider the representation

$$u_{n-2}(\rho) \equiv \int_0^\pi e^{-i\rho\cos\phi}\sin^{n-2}\phi\,d\phi = \int_0^\pi e^{-i\rho\cos\phi}(1-\cos^2\phi)\sin^{n-4}\phi\,d\phi. \tag{32}$$

Then one can easily get the recurrence formula

$$u_{n-2}(\rho) = u_{n-4}(\rho) + u_{n-4}''(\rho), \quad n \geq 4. \tag{33}$$

The function $u_0(\rho) = J_0(\rho)$ is the solution of the Bessel equation

$$u_0''(\rho) + \frac{1}{\rho}u_0'(\rho) + u_0(\rho) = 0.$$

Some known properties of this equation will be considered. The $\nu^{th}$ Bessel function $J_\nu(x)$ is the solution of the Bessel equation

$$y'' + \frac{1}{x}y' + \left(1 - \frac{\nu^2}{x^2}\right)y = 0,$$

where $\nu$ is a real or complex number, the real part of which is nonnegative. This equation has a singularity at $x=0$. For applications one needs the case $\nu = 0$. With the help of power series $J_0(x)$ can be represented as (e.g., [TS])

$$J_0(x) = \sum_{k=0}^{\infty} (-1)^k \frac{1}{[\Gamma(k+1)]^2} \left(\frac{x}{2}\right)^{2k}.$$

Moreover, one has the recurrence formula

$$J_0'(x) = -J_1(x).$$

The Bessel equation has a second linearly independent solution $K_0(x)$ defined by the Hankel function $H_0^{(1)}$

$$K_0(x) = \frac{1}{2} \pi i H_0^{(1)}(ix),$$

which is a real function for real $x$. Its integral representation is

$$K_0(x) = \int_0^\infty e^{-x \operatorname{ch} \eta} d\eta, \quad x > 0,$$

which vanishes exponentially as when $x \to \infty$ and has a logarithmic singularity for $x \to 0$

$$K_0(x) = -\ln x + \cdots,$$

where the dots refer to the bounded part for $x \to \infty$.
Then $u_1(\rho) = \frac{2 \sin \rho}{\rho}$ is the solution of the equation

$$u_1'' + \frac{2}{\rho} u_1' + u_1 = 0.$$

Thus it follows from (33) that

$$u_2(\rho) = u_0'' + u_0 = -\frac{1}{\rho} u_0'(\rho),$$

$$u_3(\rho) = u_1'' + u_1 = -\frac{2}{\rho} u_1'(\rho).$$

One can obtain that

$$u_k(\rho) = \frac{const}{\rho} u_{k-2}'(\rho),$$

is the solution of the equation

$$u_k'' + \frac{k+1}{\rho} u_k' + u_k = 0, \quad k \geq 0.$$

So (33) is represented as

$$u_{n-2}(\rho) = -\frac{n-3}{\rho} u'_{n-4}(\rho), \quad n \geq 4,$$

which gives the representation

$$u_{n-2}(\rho) = (-1)^{m-1}(2m-3)!! \left(\frac{1}{\rho}\frac{d}{d\rho}\right)^{m-1} u_0(\rho) \text{ for } n = 2m,$$
$$u_{n-2}(\rho) = (-1)^{m-1}(2m-2)!! \left(\frac{1}{\rho}\frac{d}{d\rho}\right)^{m-1} u_1(\rho) \text{ for } n = 2m+1.$$
(34)

Therefore by (32), (34), and because

$$\frac{1}{2\rho}\frac{d}{d\rho} = \frac{d}{d\rho^2},$$

we can represent (27) in the form

$$\widehat{f}(\rho) = (-2)^{m-1} \frac{d^{m-1}}{d(\rho^2)^{m-1}} \int_0^\infty f(r) r J_0(r\rho) dr, \quad \text{for } n = 2m, \quad (35)$$

$$\widehat{f}(\rho) = (-2)^m \sqrt{\frac{2}{\pi}} \frac{d^m}{d(\rho^2)^m} \int_0^\infty f(r) \cos r\rho \, dr, \quad \text{for } n = 2m+1, \ m \geq 1. \quad (36)$$

*Important examples of Fejér kernels.*
**1. Gauss–Weierstrass kernel.** Let

$$H_n(x) = \exp[-a|x|^2], \quad a > 0, \quad x \in R^n. \quad (37)$$

In the case $n = 1$ the FIT of (37) can be calculated simply by taking its derivative with respect to the parameter of the FIT and then integrating by parts. One easily obtains a linear differential equation of first order, the solution of which is represented as

$$\widehat{H}_1(y) = K_1(y) = \frac{1}{\sqrt{2a}} \exp\left[-\frac{y^2}{4a}\right].$$

Because $H_n(x) = H_1(x_1) H_1(x_2) \cdots H_1(x_n)$, by (18), for $n \geq 1$ one can get

$$\widehat{H}_n(y) = K_n(y) = \frac{1}{(\sqrt{2a})^n} \exp\left[-\frac{|y|^2}{4a}\right]. \quad (38)$$

**2. Abel–Poisson kernel.** Let

$$H_n(x) = e^{-a|x|}, \quad a > 0, \quad x \in R^n. \tag{39}$$

The FIT of (39) can be obtained simply by using (35), (36). For $n = 3$ by (36) we have

$$\frac{1}{(\sqrt{2\pi})^3} \int_{R^3} e^{-a|x|} e^{-i(x \cdot y)} dx = -\sqrt{\frac{2}{\pi}} \frac{1}{\rho} \frac{d}{d\rho} \frac{a}{a^2 + \rho^2} = -\sqrt{\frac{2}{\pi}} \frac{d}{da} \frac{1}{a^2 + \rho^2}. \tag{40}$$

Then using the inversion theorem (22) with respect to only one variable, for instance $y_3$, one can get

$$\frac{1}{2\pi} \int_{R^2} e^{-a\sqrt{r^2+x_3^2}} e^{-i(x_1 y_1 + x_2 y_2)} dx_1 dx_2 = -\frac{1}{\pi} \frac{d}{da} \int_R \frac{e^{ix_3 y_3} dy_3}{y_3^2 + b^2},$$

where $r^2 = x_1^2 + x_2^2$, $b^2 = a^2 + y_1^2 + y_2^2 = a^2 + \rho^2$. Because $F(z) = e^{ix_3 z}(z+ib)^{-1}$ with $b > 0$ is an analytic function of the complex variable $z$ in the half plane $J_m z > 0$ that, for $x_3 > 0$, vanishes at infinity, the last integral can be calculated by the Cauchy integral formula of one complex variable. The result is $2\pi i F(ib)$. Hence

$$\frac{1}{2\pi} \int_{R^2} e^{-a\sqrt{r^2+x_3^2}} e^{-i(x \cdot y)} dx = -\frac{d}{da} \frac{e^{-x_3\sqrt{a^2+\rho^2}}}{\sqrt{a^2 + \rho^2}}, \quad x_3 \geq 0. \tag{41}$$

Therefore, by (35), we have

$$\int_0^\infty e^{-a\sqrt{r^2+x_3^2}} r J_0(r\rho) dr = -\frac{d}{da} \frac{e^{-x_3\sqrt{a^2+\rho^2}}}{\sqrt{a^2 + \rho^2}}, \quad x_3 \geq 0, \quad a > 0. \tag{42}$$

In this equality let $x_3 = 0$. Then

$$\int_0^\infty e^{-ar} r J_0(r\rho) dr = -\frac{d}{da} \frac{1}{\sqrt{a^2 + \rho^2}}. \tag{43}$$

From (42) the Sommerfeld formula easily follows too

$$\int_0^\infty e^{-a\sqrt{r^2+x_3^2}} \frac{r J_0(r\rho) dr}{\sqrt{r^2 + x_3^2}} = \frac{e^{-x_3\sqrt{a^2+\rho^2}}}{\sqrt{a^2 + \rho^2}}, \quad x_3 \geq 0. \tag{44}$$

So by (35), (36), (43) one can obtain

$$K(\rho) = F_y[e^{-a|x|}] = C_n \frac{a}{(a^2 + \rho^2)^{\frac{n+1}{2}}}, \quad n \geq 1, \quad a > 0, \qquad (45)$$

where $C_n = (n-1)!!$ for $n = 2m$, and $C_n = \sqrt{\frac{2}{\pi}}(n-1)!!$ for $n = 2m+1$.

**3. Sommerfeld kernel.** Let

$$H_n(r) = e^{-a\sqrt{r^2+k^2}}, \quad a > 0, \quad r = |x|. \qquad (46)$$

It is known that

$$\frac{1}{\sqrt{2\pi}} \int_R \exp\left[-a\sqrt{x_1^2 + k^2} - ix_1 y_1\right] \frac{dx_1}{\sqrt{x_1^2 + k^2}} = \sqrt{\frac{\pi}{2}} i H_0^{(1)}(ikr_1), \qquad (47)$$

where $r_1^2 = a^2 + y_1^2$ and $H_0^{(1)}$ is the Hankel function of zeroth order. From this equality it follows that

$$\int_0^\infty \exp\left[-a\sqrt{r^2+k^2}\right] \cos r\rho \, dr = -\frac{i\pi}{2} \frac{\partial}{\partial a} H_0^{(1)}(ik\sqrt{a^2+\rho^2}). \qquad (48)$$

So by (42), (35), (36) we have

$$\begin{aligned} K(\rho) &= F_y[H_n(r)] = -i(-2)^m \sqrt{\frac{\pi}{2}} \frac{\partial}{\partial a} \frac{\partial^m}{\partial(\rho^2)^m} H_0^{(1)}(ik\sqrt{a^2+\rho^2}) \\ &\qquad \text{for } n = 2m+1, \\ K(\rho) &= F_y[H_n(r)] = -(-2)^{m-1} \frac{\partial}{\partial a} \frac{\partial^{m-1}}{\partial(\rho^2)^{m-1}} \frac{e^{-k\sqrt{a^2+\rho^2}}}{\sqrt{a^2+\rho^2}} \\ &\qquad \text{for } n = 2m. \end{aligned} \qquad (49)$$

All these kernels and their FIT will be used to solve explicitly certain B&IVP.

# 1 BVP for Regular and Generalized Regular Functions, and the Hobson Formula in Clifford Analysis

Let $u(x)$ be a regular function, i.e., a solution of (8) in the space $R_{(n)}$,

$$\bar{\partial} u = 0, \qquad (1.1)$$

or a generalized regular or $h$-regular function if

## 1. BVP for ... Hobson Formula in Clifford Analysis

$$\overline{\partial}u + \widetilde{u}h = 0, \tag{1.2}$$

where $\widetilde{u}$ is defined by (6). Then $u(x)$, $x(x_0, \ldots, x_n)$ is also a solution of the equation

$$\Delta u = 0 \quad \text{in the case (1.1)}, \tag{1.3}$$
$$\Delta u - |h|^2 u = 0 \quad \text{in the case (1.2)}, \tag{1.4}$$

supposing that $h$ is a vectorial constant $h = \sum_{k=0}^{n} h_k e_k$.

The following equalities can be easily proved.

(a) Let $u$ be a scalar function and $v \in R_{(n)}$. Then

$$\overline{\partial}(uv) = \overline{\partial}(vu) = u\overline{\partial}v + v\overline{\partial}u. \tag{1.5}$$

(b) Let $v$ be a constant. Then for any $u \in R_{(n)}$

$$\overline{\partial}(uv) = (\overline{\partial}u)v, \quad (vu)\overline{\partial} = v(u\overline{\partial}). \tag{1.6}$$

Using (1.6) one can obtain from (1.2) the equality (1.4).

First, the solution of the BVP for equation (1.3) in the half space $x_n > 0$ and in the ball $|x| < 1$ is given. For this, the well-known Gauss formula is used for any $u, v$ that have second-order continuous derivatives inside $\sigma$ and first-order derivatives on $\sigma \cup S$, where $S$ is the boundary of $\sigma$,

$$\int_\sigma (v\Delta u - u\Delta v) d\sigma = \int_S \left( v \frac{\partial u}{\partial n} - u \frac{\partial v}{\partial n} \right) dS.$$

From this it follows for any regular harmonic functions $u, v$ that

$$\int_S \left( v \frac{\partial u}{\partial n} - u \frac{\partial v}{\partial n} \right) dS = 0, \tag{1.7}$$

where $n$ is an outward unit normal of $S$.

Let $v$ be a fundamental solution of equation (1.3), $n \geq 2$. Then taking $v = \frac{1}{r^{n-1}}$, one can prove

$$\lim_{\varepsilon \to 0} \int_{S_\varepsilon} \left( \frac{1}{r^{n-1}} \frac{\partial u}{\partial n} - u \frac{\partial}{\partial n} \frac{1}{r^{n-1}} \right) dS = -\omega_n(n-1)u(x),$$

## 78    II. Multidimensional Cases

where $r^2 = |x - \xi|^2$ and $S_\varepsilon$ is a sphere with radius $\varepsilon$ and center at $x$. Thus using (1.7) one has the classical Green formula for any harmonic function $u$

$$u(x) = \frac{1}{(n-1)\omega_n} \int_S \left( \frac{1}{r^{n-1}} \frac{\partial u}{\partial n} - u \frac{\partial \frac{1}{r^{n-1}}}{\partial n} \right) dS. \tag{1.8}$$

Let $u, v$ be regular harmonic functions. Then by force of (1.7), (1.8) we have

$$u(x) = \frac{1}{(n-1)\omega_n} \int_S \left[ \left( \frac{1}{r^{n-1}} - v \right) \frac{\partial u}{\partial n} - u \frac{\partial}{\partial n} \left( \frac{1}{r^{n-1}} - v \right) \right] dS. \tag{1.9}$$

Consider the classical Dirichlet and Neumann problems. The Function

$$G(x, \xi) = \frac{1}{r^{n-1}} - v, \tag{1.10}$$

is called the Green's function for the Dirichlet problem if $v$ is a harmonic function for which

$$G(x, \xi) = 0, \quad \xi \in S, \quad x \in \sigma, \tag{1.11}$$

or for the Neumann problem if

$$\frac{\partial G}{\partial n} = 0, \quad \xi \in S, \quad x \in \sigma. \tag{1.12}$$

It is easy to define the Green's function for the half space $x_n > 0$ and for the ball $|x| < 1$.

The Green's function of the Dirichlet problem for the half space $x_n > 0$ is

$$G(x, \xi) = \frac{1}{r_{pm}^{n-1}} - \frac{1}{r_{p_1m}^{n-1}}, \quad p(\xi), \quad m(x),$$

where $p, p_1$ are symmetric points with respect to $x_n = 0$. When $p$ is on the boundary, one has $G = 0$. Thus by the Dirichlet condition the harmonic function $u(x)$ for $x_n > 0$ is defined as

$$u(x) = \frac{2x_n}{\omega_n} \int_{R^n} \frac{u(\xi) \, d\xi}{(r^2 + x_n^2)^{\frac{n+1}{2}}}, \tag{1.13}$$

where $r^2 = (x_0 - \xi_0)^2 + \cdots + (x_{n-1} - \xi_{n-1})^2$.

For the Neumann problem the Green's function is

$$G(x, \xi) = \frac{1}{r_{pm}^{n-1}} + \frac{1}{r_{p_1 m}^{n-1}}, \qquad (1.14)$$

and the solution is

$$u(x) = \frac{1}{(n-1)\omega_n} \int_{R^n} \frac{\varphi(\xi)\, d\xi}{(r^2 + x_n^2)^{\frac{n-1}{2}}}, \qquad (1.15)$$

where

$$\frac{\partial u}{\partial \xi_n} = \varphi(\xi) \quad \text{for} \quad \xi_n = 0.$$

Now the Green's function of the Dirichlet problem for the ball $|x| < 1$ is

$$G(x, \xi) = \frac{1}{r_{pm}^{n-1}} - \left(\frac{1}{\rho \cdot r_{p_1 m}}\right)^{n-1}, \quad p(\xi),\ m(x), \qquad (1.16)$$

where $p$, $p_1$ are symmetric points with respect to the sphere, $\rho_1 = |op_1|$, $\rho = |op|$, $\rho\rho_1 = 1$, $p_1(\xi')$, and $\xi' = \frac{\xi}{|\xi|^2}$.

To prove that

$$v(\xi) = \frac{1}{(|\xi|\, |x - \xi'|)^{n-1}}, \qquad (1.17)$$

is a harmonic function in the space $R^{n+1}$ as a function of $\xi$, we use a theorem due to Lord Kelvin: if $u(\xi)$ is harmonic, then $v(\xi) = \frac{1}{|\xi|^{n-1}} u(\frac{\xi}{|\xi|^2})$ is harmonic. Thus, since $\frac{1}{|x - \xi|^{n-1}}$ is harmonic, $v(\xi)$ defined by (1.17) is harmonic too, and $G(x, \xi)$ defined by (1.16) is zero when $|\xi| = 1$.

To obtain the solution of the Dirichlet problem for $|\xi| = 1$ use $\frac{\partial G}{\partial n_\xi} = \frac{\partial G}{\partial |\xi|}$. The symbols $\xi$, $x$ denote points as well as radius vectors, so scalar products define lengths of vectors,

$$|\xi - x|^2 = (\xi - x)(\xi - x) = |\xi|^2 - 2|x|\,|\xi|\cos(x\xi) + |x|^2,$$
$$|\xi|^2\,|\xi' - x|^2 = 1 - 2|x|\,|\xi|\cos(x\xi) + |\xi|^2|x|^2.$$

Therefore, for $|\xi| = 1$

$$\frac{\partial}{\partial |\xi|} \frac{1}{|\xi - x|^{n-1}} = \frac{1-n}{|\xi - x|^{(n+1)}} (1 - |x|\cos(x\xi)),$$
$$\frac{\partial}{\partial |\xi|} \frac{1}{(|\xi|\,|\xi - x|)^{n-1}} = \frac{1-n}{|\xi - x|^{(n+1)}} (|x|^2 - |x|\cos(x\xi)).$$

## II. Multidimensional Cases

Thus

$$\frac{\partial G}{\partial |\xi|} = \frac{(1-n)(1-|x|^2)}{|\xi - x|^{(n+1)}},$$

and by (1.8), one has the Poisson formula

$$u(x) = \frac{1}{\omega_n} \int_{|\xi|=1} \frac{(1-|x|^2)u(\xi)\,dS_\xi}{|\xi - x|^{(n+1)}}. \tag{1.18}$$

Now consider the Neumann problem for the ball $|x| < 1$. Define the harmonic function $u(x)$ in $|x| < 1$ by the condition

$$\frac{\partial u}{\partial \rho} = f(x), \quad |x| = \rho = 1. \tag{1.18_1}$$

**Solution.** It is easy to see that $r\frac{\partial u}{\partial r}$ is harmonic in $|x| < 1$. Therefore, by force of (1.18), (1.18$_1$) it is represented as

$$r\frac{\partial u}{\partial r} = \frac{1}{\omega_n} \int_{|\xi|=1} \frac{f(\xi)(1-|x|^2)}{|\xi - x|^{(n+1)}}\,dS_\xi, \quad r = |x|. \tag{1.19_1}$$

Moreover, for the Neumann problem we have

$$\int_{|\xi|=1} f(\xi)\,dS_\xi = 0.$$

Consider the case $n = 2$ and define $u(x)$ from (1.19$_1$)

$$u(x) = \frac{1}{4\pi} \int_{|\xi|=1} f(\xi)\,dS_\xi \int \frac{(1-r^2)\,dr}{r[1+r^2-2r\cos(x\xi)]^{3/2}}.$$

One can obtain

$$\int \frac{(1-r^2)\,dr}{r[1+r^2-2r\cos(x\xi)]^{3/2}} = \int \frac{dr}{r[1+r^2-2r\cos(x\xi)]^{1/2}} -$$
$$- \int \frac{[2r-2\cos(x\xi)]\,dr}{[1+r^2-2r\cos(x\xi)]^{3/2}} =$$
$$= \int \frac{dr}{r^2[1+\frac{1}{r^2}-\frac{2}{r}\cos(x\xi)]^{1/2}} + \frac{2}{[1+r^2-2r\cos(x\xi)]^{1/2}} =$$
$$= -\ln[1-r\cos(x\xi)+|x-\xi|] + \frac{2}{|x-\xi|} + \ln r.$$

Thus the solution of the Neumann problem is represented

$$u(x) = -\frac{1}{4\pi} \int_{|\xi|=1} f(\xi)\left\{\ln[1 - r\cos(x\xi) + |x - \xi|] - \frac{2}{|x - \xi|}\right\} dS_\xi. \tag{1.20$_1$}$$

Note that for $n \geq 3$ the integral is

$$\int \frac{(1 - r^2)\, dr}{r[1 + r^2 - 2r\cos(x\xi)]^{\frac{n+1}{2}}}.$$

For $n = 2m + 1$ the integral can be easily defined and for $n = 2m$ it is defined as for $n = 2$.

Thus the Neumann problem for a harmonic function in the ball is solved in quadratures.

The Riemann–Schwartz principle of reflection can be used successfully to solve BVP for harmonic functions in the multidimensional case too.

(a) Let $u(x)$ be a harmonic function in $D$ the boundary of which contains $S$, the part of the plane $x_n = 0$, where

$$u(x_0, \ldots, x_{n-1}, 0) = 0 \quad \text{or} \quad \frac{\partial u}{\partial x_n} = 0, \quad x_n = 0.$$

Then in the first case the function

$$U(x) = \begin{cases} u(x), & x_n > 0, \\ -u(x_0, \ldots, x_{n-1}, -x_n), & x_n < 0, \end{cases} \tag{1.19}$$

and in the second case

$$U(x) = \begin{cases} u(x), & x_n > 0, \\ u(x_0, \ldots, x_{n-1}, -x_n), & x_n < 0, \end{cases}$$

is a harmonic function in $D \cup D^* \cup S$, where $D^*$ is a reflection of $D$ with respect to $x_n = 0$.

(b) Let the boundary of $D$ contain $S$, the part of the sphere $|x| = 1$. If for the harmonic function in $D$ one has

$$u(x) = 0, \quad x \in S, \quad \text{or} \quad \frac{\partial u}{\partial r} = 0, \quad r = 1,$$

then in the first case

$$U(x) = \begin{cases} u(x), \\ |x| < 1, \\ -\frac{1}{|x|^{n-1}} u(\frac{x}{|x|^2}), \\ |x| > 1, \end{cases} \tag{1.20}$$

and in the second case

$$U(x) = \begin{cases} u(x), \\ |x| < 1, \\ \frac{1}{|x|^{n-1}} u(\frac{x}{|x|^2}), \\ |x| > 1, \end{cases}$$

is a harmonic in $D \cup D^* \cup S$, where $D^*$ is a reflection of $D$ with respect to $|x| = 1$.

**Problems for the half ball.** Find in the half ball $|x| \leq 1$, $x_n > 0$, a regular harmonic function when the Dirichlet condition is given on the spherical part

$$u(x) = \varphi(x), \quad |x| = 1, \quad x_n > 0,$$

and on $x_n = 0$

$$u(x) = \psi(x) \quad \text{or} \quad \frac{\partial u}{\partial x_n} = \psi(x).$$

**Solution.** Represent $u(x)$ as

$$u(x) = u_1(x) + u_2(x),$$

where $u_1, u_2$ are harmonic functions satisfying the conditions

$$\begin{aligned} u_1|_S = \varphi(x), \quad x \in S, \quad & \frac{\partial u_1}{\partial x_n} = 0 \quad \text{or} \quad u_1 = 0 \text{ for } x_n = 0, \\ u_2|_S = 0, \quad & \frac{\partial u_2}{\partial x_n} = \psi(x) \quad \text{or} \quad u_2 = \psi \text{ for } x_n = 0, \end{aligned} \tag{1.21}$$

where $S$ is the surface of the half sphere $|x| = 1$, $x_n > 0$. By force of (1.19), (1.20) the functions

$$v_1(x) = \begin{cases} u_1(x), \\ |x| \leq 1, \quad x_n > 0, \\ \pm u_1(x_0, \ldots, x_{n-1}, -x_n), \\ |x| \leq 1, \quad x_n < 0, \end{cases} \quad (1.22)$$

$$v_2(x) = \begin{cases} u_2(x), \\ |x| \leq 1, \quad x_n > 0, \\ -\frac{1}{|x|^{n-1}} u_2(\frac{x}{|x|^2}), \\ |x| \geq 1, \quad x_n > 0, \end{cases} \quad (1.23)$$

where the sign $(+)$ in (1.22) applies when $\frac{\partial u_1}{\partial x_n} = 0$ and $(-)$ when $u_1 = 0$ for $x_n = 0$, are holomorphic correspondingly in the ball $|x| < 1$ and in the half space $x_n > 0$. By force of (1.21) $v_1$ satisfies the Dirichlet condition for the ball. Thus $v_1$ is defined by (1.18). Similarly, $v_2$ satisfies the Neumann or Dirichlet condition for the half space $x_n > 0$; i.e., $v_2$ is defined by (1.15) or (1.13). Note that in the case of the Neumann condition for $u_2$ in (1.21) we use $\frac{\partial |x|}{\partial x_n} = 0$, $x_n = 0$, and

$$\frac{\partial v_2}{\partial x_n} = \begin{cases} \psi(x), \\ |x| \leq 1, \quad x_n = 0, \\ -\frac{1}{|x|^{n+1}} \psi(\frac{x}{|x|^2}), \\ |x| \geq 1, \quad x_n = 0. \end{cases}$$

In a perfectly analogous way the solutions of the problems for the domain $x_n \geq 0$, $x_{n-1} \geq 0$, $(x_0, \ldots, x_{n-2}) \in R^{n-1}$ can be defined with the Dirichlet, Neumann or mixed conditions so that they reduce to corresponding problems for the half space using (1.15) or (1.13).

All the above problems can be solved for the Poisson equation $\Delta u = F(x)$ using its partial solution

$$u(x) = \int_\sigma G(x, \xi) F(\xi) \, d\xi, \quad (1.24)$$

where $G$ is the Green's function of the corresponding problem.

**Dirichlet and Neumann problem for the Helmholtz equation.**

$$\Delta u - h^2 u = 0, \quad h = const. \quad (1.25)$$

## II. Multidimensional Cases

Let $R_+^{n+1}$ be the half space $x_n > 0$, $x(x_0, \ldots, x_{n-1})$, and let $u(x, x_n)$ be a real function. In this space find the solution of (1.25) that vanishes at infinity and satisfies the conditions

$$u(x, 0) = f(x) \text{ for the Dirichlet problem,} \tag{1.26}$$

$$\frac{\partial u}{\partial x_n} = \phi(x) \text{ for the Neumann problem,} \tag{1.27}$$

on the boundary $x_n = 0$ where the given functions $f, \phi \in L(R^n)$. The solution is supposed to be in the class of functions satisfying the conditions:

(a) $u(x, x_n)$, $\dfrac{\partial u}{\partial x_k}$, $\dfrac{\partial^2 u}{\partial x_k^2}$, $k = 0, 1, \ldots, n-1$, belong to $L(R^n)$ for $\forall x_n \geq 0$,

(b) $\left|\dfrac{\partial u}{\partial x_n}\right| \leq f_1(x)$, $\left|\dfrac{\partial^2 u}{\partial x_n^2}\right| \leq f_2(x)$, where $f_1, f_2 \in L(R^n)$.

**Solution.** By the FIT of equation (1.25) with respect to the variables $x_0, x_1, \ldots, x_{n-1}$, one can get

$$\frac{d^2 \widehat{u}(y, x_n)}{dx_n^2} - (|y|^2 + h^2)\widehat{u}(y, x_n) = 0, \quad y \in R^n,$$

and by (1.26), (1.27)

$$\widehat{u}(y, 0) = \widehat{f}(y) \quad \text{or} \quad \frac{d\widehat{u}(y, x_n)}{dx_n} = \widehat{\phi}(y) \text{ for } x_n = 0.$$

The solution that vanishes at infinity of this simple problem for ordinary differential equations is represented as

$$\widehat{u}(y, x_n) = \widehat{f}(y) e^{-x_n \lambda} \quad \text{or} \quad \widehat{u}(y, x_n) = -\frac{1}{\lambda} \widehat{\phi}(y) e^{-x_n \lambda}, \quad x_n > 0,$$

where $\lambda = \sqrt{|y|^2 + h^2}$. Because these functions belong to $L(R^n)$ for $x_n > 0$, the inversion formula (22) can be used. Thus

$$u(x, x_n) = \frac{1}{(\sqrt{2\pi})^n} \int_{R^n} \widehat{f}(y) e^{-\lambda x_n} e^{i(x \cdot y)} dy, \tag{1.28}$$

or

$$u(x, x_n) = \frac{-1}{(\sqrt{2\pi})^n} \int_{R^n} \widehat{\phi}(y) e^{-\lambda x_n} e^{i(x \cdot y)} \frac{dy}{\lambda}. \tag{1.29}$$

Using (49) these can be written in the form

$$u(x, x_n) = \frac{(-1)^{m+1}}{\pi^m h} \frac{\partial}{\partial x_n} \frac{\partial^m}{\partial (x_n^2)^m} \int_{R^n} f(y) e^{-|h|r} dy \quad \text{for } n = 2m,$$

$$u(x, x_n) = \frac{(-1)^{m+1} i}{2\pi^m} \frac{\partial}{\partial x_n} \frac{\partial^m}{\partial (x_n^2)^m} \int_{R^n} f(y) H_0^{(1)}(i|h|r) dy \quad \text{for } n = 2m + 1.$$
(1.30)

The solution of the Neumann problem is

$$u(x, x_n) = \frac{(-1)^{m+1}}{\pi^m h} \frac{\partial^m}{\partial (x_n^2)^m} \int_{R^n} \phi(y) e^{-|h|r} dy \quad \text{for } n = 2m,$$

$$u(x, x_n) = \frac{(-1)^{m+1} i}{2\pi^m} \frac{\partial^m}{\partial (x_n^2)^m} \int_{R^n} \phi(y) H_0^{(1)}(i|h|r) dy \quad \text{for } n = 2m + 1,$$
(1.31)

where $r^2 = |x - y|^2 + x_n^2$. That they are the solutions of above problems can be directly verified.

From these representations remarkable properties follow: if the given functions $f, \phi$ are odd with respect to some fixed variable $x_k$ ($0 \leq k \leq n - 1$), then

$$u(x, x_n) = 0 \quad \text{for } x_k = 0, \quad x_n > 0,$$

and if they are even, then

$$\frac{\partial u}{\partial x_k} = 0, \quad x_k = 0, \quad x_n > 0.$$

As we saw above in the two-dimensional case for holomorphic functions, in some planar domain with cuts many problems are solved. In the high-dimensional case we have no such solution. We will consider a three-dimensional space with a crack along the circular domain $x_0^2 + x_1^2 \leq a^2$, and some problems for harmonic functions will be solved.

Let $\Omega_+$ be the half space $x_2 > 0$, $x(x_0, x_1, x_2) \in \Omega_+$. On the boundary $x_2 = 0$ the domain $x_0^2 + x_1^2 < a^2$ is denoted by $S_+$ and the domain $x_0^2 + x_1^2 > a^2$ by $S_-$, $a = const$. Let $\Omega$ be the infinite space $R^3$ with a crack along $S_+$. To solve the corresponding problems we will use the solution of a problem for harmonic functions constructed by Hobson [Ho]. First Hobson's representation will be given.

**Problem** (Hobson). Find a harmonic function $u(x)$ in $\Omega$ that vanishes at infinity satisfying the conditions

$$u^{\pm}(x_0, x_1, 0) = f(x_0, x_1) \quad \text{on } S_+, \quad (1.32)$$

## II. Multidimensional Cases

where

$$u^{\pm}(x_0, x_1, 0) = \lim_{x_2 \to \pm 0} u(x_0, x_1, x_2).$$

The function is here and everywhere sufficiently smooth.

The solution is represented by Hobson's formula

$$u(x_0, x_1, x_2) = \iint_{S_+} f(\xi, \eta) \left[ \frac{x_2}{\pi^2 r^3} \left( \frac{1}{M} + \text{arctg } M \right) \right] d\xi \, d\eta, \quad (1.33)$$

where

$$M = \frac{\sqrt{2} x_2 \sqrt{a^2 - \xi^2 - \eta^2}}{r \sqrt{x_0^2 + x_1^2 + x_2^2 - a^2 + R}}, \quad (1.34)$$

$$r^2 = (x_0 - \xi)^2 + (x_1 - \eta)^2 + x_2^2, \quad R^2 = (a^2 - x_0^2 - x_1^2 - x_2^2)^2 + 4a^2 x_2^2.$$

It is clear that when $x_2 = 0$ and $x_0^2 + x_1^2 \geq a^2$, then $R = x_0^2 + x_1^2 - a^2$, and if $x_0^2 + x_1^2 \leq a^2$, then $R = a^2 - x_0^2 - x_1^2$.

If we have the boundary conditions

$$u^+(x_0, x_1, 0) = f_1(x_0, x_1), \quad u^-(x_0, x_1, 0) = f_2(x_0, x_1), \quad (1.35)$$

then one can consider the two harmonic functions

$$u_1(x_0, x_1, x_2) = \frac{1}{2}[u(x_0, x_1, x_2) + u(x_0, x_1, -x_2)],$$
$$u_2(x_0, x_1, x_2) = \frac{1}{2}[u(x_0, x_1, x_2) - u(x_0, x_1, -x_2)], \quad (1.36)$$

in $\Omega$ for which by (1.35) one has the boundary conditions on $S_+$

$$u_1^+ = u_1^- = \frac{1}{2}(f_1 + f_2) \equiv f(x_0, x_1),$$
$$u_2^+ = -u_2^- = \frac{1}{2}(f_1 - f_2) \equiv g(x_0, x_1).$$

Thus $u_1$ is represented by (1.33) and $u_2$ by Poisson's formula

$$u_2 = \frac{x_2}{2\pi} \iint_{S_+} \frac{g(\xi, \eta) \, d\xi \, d\eta}{r^3}. \quad (1.37)$$

**Problem 1.** Find a harmonic function $u(x)$ in $\Omega_+$ that vanishes at infinity satisfying the conditions

$$u(x_0, x_1, 0) = f(x_0, x_1) \quad \text{for} \quad (x_0, x_1) \in S_+, \tag{1.38}$$

$$\frac{\partial u}{\partial x_2} = g(x_0, x_1) \quad \text{for} \quad x_2 = 0, \quad (x_0, x_1) \in S_-. \tag{1.39}$$

Using the solution of the Neumann problem in a half space one can reduce (1.39) to the homogeneous condition $g \equiv 0$. Then, by the Schwartz reflection principle the function

$$u_1(x_0, x_1, x_2) = \begin{cases} u(x_0, x_1, x_2) \\ \quad \text{for } x_2 > 0, \\ u(x_0, x_1, -x_2) \\ \quad \text{for } x_2 < 0, \end{cases} \tag{1.40}$$

is harmonic in $\Omega$ which by (1.38) satisfies

$$u_1^+(x_0, x_1, 0) = u_1^-(x_0, x_1, 0) = f(x_0, x_1). \tag{1.41}$$

on $S_+$. Thus $u_1$ is represented by the Hobson formula (1.33).

Let $D$ be the quarter space $x_1 \geq 0$, $x_2 \geq 0$, $-\infty \leq x_0 \leq \infty$. Find the harmonic function in $D$ $u(x)$ that vanishes at infinity by the conditions

$$\frac{\partial u}{\partial x_2} = f(x_0, x_1) \quad \text{for} \quad S_-, \quad u(x_0, x_1, 0) = 0 \quad \text{for} \quad S_+, \tag{1.42}$$

$$u(x_0, 0, x_2) = 0 \quad \text{or} \quad \frac{\partial u}{\partial x_1} = 0 \quad \text{for} \quad x_1 = 0, \quad x_2 \geq 0. \tag{1.43}$$

By the Schwartz principle of reflection this problem can be easily reduced to the problem with conditions (1.41). Thus the solution can be represented like (1.33).

**Problem 2.** Let $D$ be the half space $x_2 > 0$ with a crack along the half circle $x_0^2 + x_2^2 \leq a^2$. Find the harmonic function in $D$ that vanishes at infinity by the boundary conditions

$$u(x_0, x_1, 0) = 0, \quad (x_0, x_1) \in R^2,$$
$$u^\pm(x_0, 0, x_2) = f(x_0, x_2), \quad x_0^2 + x_2^2 \leq a^2.$$

**Solution.** The harmonic function in the half space with crack along $x_0^2 + x_2^2 \leq a^2$

88    II. Multidimensional Cases

$$U(x) = \begin{cases} u(x), & x_2 > 0, \\ -u(x_0, x_1, -x_2), & x_2 < 0, \end{cases}$$

satisfies the conditions

$$U^{\pm}(x_0, 0, x_2) = \begin{cases} f(x_0, x_2), & x_2 > 0, \; x_0^2 + x_2^2 \le a^2, \\ -f(x_0, x_2), & x_2 < 0, \; x_0^2 + x_2^2 \le a^2. \end{cases}$$

Again using Hobson's representation (1.33) the solution can be written in quadratures.

The problems solved above for harmonic functions will be used below to solve corresponding problems for (1.1).

Consider the space $R_{(2)}$, $u(x) = u_0 e_0 - u_1 e_1 - u_2 e_2 - u_{12} e_1 e_2$. By force of (2), (8) equation (1.1) is a Moisil–Theodorescu system

$$\operatorname{div} U = 0, \quad \operatorname{grad} u_{12} + \operatorname{rot} U = 0, \quad U(u_0, u_1, u_2),$$

i.e.,

$$\begin{aligned}
\frac{\partial u_0}{\partial x_0} + \frac{\partial u_1}{\partial x_1} + \frac{\partial u_2}{\partial x_2} &= 0, \\
\frac{\partial u_{12}}{\partial x_0} + \frac{\partial u_2}{\partial x_1} - \frac{\partial u_1}{\partial x_2} &= 0, \\
\frac{\partial u_{12}}{\partial x_1} + \frac{\partial u_0}{\partial x_2} - \frac{\partial u_2}{\partial x_0} &= 0, \\
\frac{\partial u_{12}}{\partial x_2} + \frac{\partial u_1}{\partial x_0} - \frac{\partial u_0}{\partial x_1} &= 0.
\end{aligned} \qquad (1.44)$$

Note that for holomorphic functions $w = \alpha + i\beta$ of one complex variable, functions $\alpha$, $\beta$ are harmonic. Conversely if $\alpha$, $\beta$ are any two harmonic functions that satisfy only one equation of the Cauchy–Riemann system, then they also satisfy the second equation if it is satisfied in one point; i.e., $w$ is a holomorphic function.

An analogous property can be easily proved for the solutions of (1.44). If $u$ is solution of (1.44), then by force of (1.3) $u_0, u_1, u_2, u_{12}$ are harmonic functions. Conversely, if they are harmonic and satisfy only two equations from (1.44), then the other two are also satisfied, if some vanishing conditions at one point are assumed. For instance, in an infinite domain these harmonic functions are supposed to be zero at infinity. Let these four functions be harmonic and satisfy the first two equations of (1.44). Then differentiate the first of the two with respect to $x_1$. Substitute $\dfrac{\partial^2 u_1}{\partial x_1^2} = -\dfrac{\partial^2 u_1}{\partial x_0^2} - \dfrac{\partial^2 u_1}{\partial x_2^2}$, into it to get

$$\frac{\partial}{\partial x_0}\left(\frac{\partial u_0}{\partial x_1} - \frac{\partial u_1}{\partial x_0}\right) + \frac{\partial}{\partial x_2}\left(\frac{\partial u_2}{\partial x_1} - \frac{\partial u_1}{\partial x_2}\right) = 0.$$

By force of the second equation we obtain

$$\frac{\partial}{\partial x_0}\left(\frac{\partial u_{12}}{\partial x_1} + \frac{\partial u_1}{\partial x_0} - \frac{\partial u_0}{\partial x_1}\right) = 0.$$

Thus we have the last equation if it is satisfied at one point. In the same way if the first equation is differentiated with respect to $x_2$, one obtains the third equation.

Now let the last two equations be satisfied. Then differentiate the last equation with respect to $x_2$ and the third equation with respect to $x_1$. After summing them one has, since $u_{12}$ is harmonic,

$$\frac{\partial}{\partial x_0}\left(\frac{\partial u_{12}}{\partial x_0} + \frac{\partial u_2}{\partial x_1} - \frac{\partial u_1}{\partial x_2}\right) = 0,$$

i.e., the second equation from (1.44). To obtain the first equation, the last equation must be differentiated with respect to $x_1$ and the third with respect to $x_2$. Then, by subtracting and taking into consideration that $u_0$ is harmonic, we can easily obtain the first equation too.

Using this property to solve the BVP for (1.44) we define four harmonic functions that satisfy only two equations. Then those functions are solutions of the given BVP.

Let $D_1$ be the half space $x_2 > 0$ and $D_2$ be the half space $x_2 > 0$ with a crack along the half circle $x_1^2 + x_2^2 < a^2$. $S_+$ denotes the circular domain $x_0^2 + x_1^2 \leq a^2$, $x_2 = 0$, and $S_-$ the domain $x_0^2 + x_1^2 \geq a^2$, $x_2 = 0$. For these domains several BVP are solved for harmonic functions. They will be used to solve following.

**Problem 3.** Find the regular solution of (1.44) in $D_1$ that vanishes at infinity by the conditions on $x_2 = 0$

$$u_0(x_0, x_1, 0) = f_0(x_0, x_1), \quad u_1(x_0, x_1, 0) = f_1(x_0, x_1). \tag{1.45}$$

**Solution.** Because $u_0$, $u_1$ are harmonic functions, by these conditions, they are represented by (1.13). By substituting for $u_0$, $u_1$ in the first and last equations of (1.44) one can define $u_2$ and $u_{12}$.

Thus two conditions in $D_1$ define all unknown functions.

**Problem 4.** Find the solution of (1.44) in $D_1$ that vanishes at infinity by the conditions

$$u_2(x_0, x_1, 0) = f(x_0, x_1), \quad u_{12}(x_0, x_1, 0) = \varphi(x_0, x_1) \quad \text{on } S_+,$$
$$\frac{\partial u_2}{\partial x_2} = f_1(x_0, x_1), \quad \frac{\partial u_{12}}{\partial x_2} = \varphi_1(x_0, x_1) \quad \text{on } S_-.$$

The solution is represented using the solution of the problem posed by (1.38), (1.39). In other words, the harmonic functions $u_2$ and $u_{12}$ are known. Then $u_0$, $u_1$ can be defined by the first and last equations of (1.44). Thus all the problems considered above for harmonic functions can be solved successfully for (1.44) when two unknown functions are given on the plane boundary.

Now consider the generalized regular equation (1.2) in the space $R_{(2)}$, i.e., the generalized Moisil–Theodorescu system

$$\text{div } U + (A \cdot U) = 0, \quad \text{grad } u_{12} + \text{rot } U + [U \times A] + A u_{12} = 0,$$

i.e.,

$$\frac{\partial u_0}{\partial x_0} + \frac{\partial u_1}{\partial x_1} + \frac{\partial u_2}{\partial x_2} + a u_0 + b u_1 + c u_2 = 0,$$
$$\frac{\partial u_{12}}{\partial x_0} + \frac{\partial u_2}{\partial x_1} - \frac{\partial u_1}{\partial x_2} + c u_1 - b u_2 + a u_{12} = 0,$$
$$\frac{\partial u_{12}}{\partial x_1} + \frac{\partial u_0}{\partial x_2} - \frac{\partial u_2}{\partial x_0} + a u_2 - c u_0 + b u_{12} = 0,$$
$$\frac{\partial u_{12}}{\partial x_2} + \frac{\partial u_1}{\partial x_0} - \frac{\partial u_0}{\partial x_1} + b u_0 - a u_1 + c u_{12} = 0.$$
(1.46)

All components of $u$ satisfy equation (1.4)

$$\Delta u - (a^2 + b^2 + c^2) u = 0. \tag{1.47}$$

We will prove that in this case as for the Moisil–Theodorescu system, if $u_k$ ($k = 0, 1, 2, 12$) are solutions of (1.47) and satisfy two equations from (1.46), then if some vanishing conditions at one point are supposed, the other two are also satisfied. For instance, let the first two equations be satisfied. Then differentiate the first with respect to $x_1$ and substitute for $\dfrac{\partial^2 u_1}{\partial x_1^2}$ using (1.47). Taking into consideration the second equation, one has

$$\frac{\partial}{\partial x_0}\left(\frac{\partial u_0}{\partial x_1} - \frac{\partial u_1}{\partial x_0} - \frac{\partial u_{12}}{\partial x_2}\right) + \frac{\partial}{\partial x_2}(-u_1 c + u_2 b - a u_{12}) +$$
$$+ a\left(\frac{\partial u_0}{\partial x_0} + a u_1\right) + b\left(\frac{\partial u_1}{\partial x_1} + b u_1\right) + c\left(\frac{\partial u_2}{\partial x_1} + c u_1\right) = 0.$$

Define $\dfrac{\partial u_1}{\partial x_1} + bu_1$ from the first equation and substitute it into this equation. At last one obtains

$$\frac{\partial}{\partial x_0}\left(\frac{\partial u_0}{\partial x_1} - \frac{\partial u_1}{\partial x_0} - \frac{\partial u_{12}}{\partial x_2} - bu_0 + au_1 - cu_{12}\right) +$$
$$+ a\left(\frac{\partial u_0}{\partial x_1} - \frac{\partial u_1}{\partial x_0} - \frac{\partial u_{12}}{\partial x_2} - bu_0 + au_1 - cu_{12}\right) = 0.$$

So we have the last equation of (1.46). Similarly we can obtain the third equation.

**Problem 5.** Find the regular solution of (1.46) in the half space $x_2 > 0$ that vanishes at infinity by the conditions (like the Dirichlet conditions)

$$u_0(x_0, x_1, 0) = f(x_0, x_1), \quad u_1(x_0, x_1, 0) = \varphi(x_0, x_1).$$

**Solution.** Because $u_0, u_1$ are solutions of (1.47) they are represented in the form (1.30). Then using the above property, it is sufficient to satisfy two equations from (1.46). In this case from the first and last equations we can define $u_2$ and $u_{12}$.

In the same way the Neumann problem can be solved using (1.31). Thus all the problems we can solve for the Helmholtz equation (1.47) can be solved for (1.46) too.

Now we consider the Riesz and generalized Riesz systems that can be obtained from (1.1), (1.2). For $n = 2$ and

$$u = u_0 e_0 + u_1 e_1 + u_2 e_2,$$

one has the overdetermined system

$$\begin{aligned} \operatorname{div} U &= 0, \quad \operatorname{div} U + (A \cdot U) = 0, \\ \operatorname{rot} U &= 0 \text{ or } \operatorname{rot} U + [U \times A] = 0, \end{aligned} \quad (1.48)$$

with four equations and three unknowns.

The regular solution of (1.48) in the half space $x_2 > 0$ that vanishes at infinity by the condition given above for harmonic functions or for the Helmholtz equation can be represented in quadratures. If $u_0$ is given, then $u_1, u_2$ are defined by the two equations of (1.48) and satisfy the other two equations too.

Let $n \geq 3$ and $u = \sum_0^n u_k e_k$. On the boundary $x_n = 0$, one $u_k$ ($0 \leq k \leq n$, $k$ is fixed) satisfies the conditions considered above for harmonic functions. The Riesz system has a unique solution that vanishes at infinity and which is represented in quadratures. For equation (1.1), when $u = \sum_A u_A e_A$, we have $2^{n-1}$ conditions for $2^n$ unknowns with $2^n$ equations, and for $x_n > 0$, the solution that vanishes at infinity can be represented explicitly.

For equations (1.1), (1.2) the Hilbert (that is, the conjugation) problem can be considered. To solve it we need the corresponding Cauchy kernels and representations in $R_{(n)}$. For regular functions in $R_{(n)}$ ($n > 1$), the $2^n$ linearly independent fundamental solutions of equation (1.1) can be written as

$$u^A = (\partial \phi) e_A, \quad A : (\alpha_1, \ldots, \alpha_k), \quad 0 \leq \alpha_1 < \ldots < \alpha_k \leq n,$$

where $\phi = |x|^{1-n}$ is a fundamental solution of the Laplace equation in $R^{n+1}$ ($n \geq 2$), $x = \sum_0^n x_k e_k$, and the Cauchy kernel is defined correspondingly as

$$K(x) = \frac{\overline{x}}{|x|^{n+1}} = \partial \frac{1}{|x|^{n-1}}.$$

To write the Cauchy kernel explicitly for equation (1.2) with $h$ a constant in $R_{(n)}$ we prove the following lemma [Ob1].

**Lemma.** *Let $g_{k-2}(r)$ be the solution of the Helmholtz equation in the space $R^{k-2}$ depending only on $r = |x|$, i.e.,*

$$\frac{d^2 g_{k-2}}{dr^2} + \frac{k-3}{r} \frac{dg_{k-2}}{dr} - |h|^2 g_{k-2}(r) = 0, \quad k \geq 4. \tag{1.49}$$

*Then the solution of the Helmholtz equation in $R^k$ depending only on $r$ is*

$$g_k(r) = \frac{1}{r} \frac{dg_{k-2}}{dr}. \tag{1.50}$$

*Proof.* From (1.49), (1.50) it is easy to obtain

$$\frac{d^2 g_k}{dr^2} + \frac{k-1}{r} \frac{dg_k}{dr} - |h|^2 g_{k-2}(r) = 0.$$

As is well known, the fundamental solution of the Helmholtz equation in $R^2$ is the zeroth-order Hankel function

$$g_2(r) = H_0^{(1)}(i|h|r), \tag{1.51}$$

and in $R^3$, the function

$$g_3(r) = \frac{1}{r} e^{-|h|r}. \tag{1.52}$$

Hence the fundamental solution in $R^{n+1}$ by (1.50) is

$$g_{n+1}(r) = C_n \left(\frac{1}{r}\frac{d}{dr}\right)^{m-1} g(r), \tag{1.53}$$

where $g(r) = g_2(r)$ when $n = 2m - 1$ and $g(r) = g_3(r)$ when $n = 2m$. It is obvious that $C_n = const$ can be chosen in such a way that

$$\lim_{r \to 0} r^{n-1} g_{n+1}(r) = 1, \quad n > 1,$$

because of the asymptotic behavior as $r \to 0$ of

$$H_0^{(1)}(i|h|r) = \frac{2i}{\pi} \ln r + \ldots,$$

where the dots indicate the finite part for $r \to 0$. Apart from this the asymptotic behavior as $r \to \infty$ has the form

$$H_0^{(1)}(ir) = -i\sqrt{\frac{2}{\pi r}}[1 + O(r^{-1})]e^{-r} \quad \text{for} \quad r \to \infty.$$

Note that because

$$\frac{1}{r}\frac{dg}{dr} = 2\frac{dg}{dr^2},$$

(1.53) can be written in the form

$$g_{n+1}(r) = C_n \frac{d^{m-1}g(r)}{d(r^2)^{m-1}}. \tag{1.54}$$

**Theorem.** *Equation* (1.2) *has $2^n$ linearly independent fundamental solutions represented as*

$$g^A(r) = (\partial g_{n+1})e_A - g_{n+1}(r)\tilde{e}_A h, \quad A(\alpha_1, \alpha_2, \ldots, \alpha_k). \tag{1.55}$$

Substituting (1.55) into equation (1.2) demonstrates that the theorem is true for all $A: 0 \leq \alpha_1 < \alpha_2 < \ldots < \alpha_k \leq n$.
For $r = |x - y|$, $x, y \in R^{n+1}$ one can get

$$\partial_x g_{n+1}(r) = \frac{\overline{x} - \overline{y}}{r^{n+1}} + O\left(\frac{1}{r^{n-1}}\right) = -\partial_y g_{n+1}(r). \qquad (1.56)$$

The expressions $\partial_x g(r)$ and $\partial_y g(r)$ denote $\partial g(r)$ where derivatives are taken with respect to the coordinates of $x$ and $y$ respectively.

Let $\Omega$ be a bounded domain in $R^{n+1}$ with a closed piecewise-smooth Lyapunov surface $S$, $u(x) : \Omega \to R_{(n)}$. Consider the two operators

$$Lu \equiv \overline{\partial} u + \widetilde{u} h, \qquad (1.57)$$
$$L_* u \equiv u \overline{\partial} - \overline{h} \widetilde{u}, \qquad (1.58)$$

which according to Lagrange are adjoint; i.e., for any $u(x), v(x) \in C^1(\Omega)$ with values in $R_{(n)}$ the following equality is true

$$\mathrm{Re}[v(Lu) + (L_* v)u] = \mathrm{Re} \sum_{A,B} \sum_{j=0}^{n} \frac{\partial v_A u_B}{\partial x_j} e_A e_j e_B, \qquad (1.59)$$

where $u = \sum_A u_A e_A$, $v = \sum_A v_A e_A$. Here we use the equality

$$\mathrm{Re}[v \widetilde{u} h] = \mathrm{Re}[\overline{h} \widetilde{v} u],$$

which one can check directly. The adjoint equation

$$v \overline{\partial} - \overline{h} \widetilde{v} = 0, \qquad (1.60)$$

has $2^n$ linearly independent fundamental solutions represented as

$$q^A = e_A(\partial g_{n+1}) + \overline{h} \widetilde{e}_A g_{n+1}, \quad A : (\alpha_1, \ldots, \alpha_k), \quad 0 \le \alpha_1 < \ldots < \alpha_k \le n, \qquad (1.61)$$

which can be also checked directly.

Let $u, v$ have continuous first-order derivatives in $\Omega$ and be continuous in the closed domain $\overline{\Omega}$, i.e., $u, v \in C^1(\Omega) \cap C(\overline{\Omega})$. Then using Stokes' well-known formula from (1.59) one can get

$$\mathrm{Re} \int_{\Omega} [v(Lu) + (L_* v)u] d\Omega = \mathrm{Re} \int_S v(y) n u(y) ds_y, \qquad (1.62)$$

where $n = \sum_0^n n_j e_j$ is defined by the outward-pointing unit normal vector to $S$ at $y \in S$, $d\Omega = dx_0 dx_1 \ldots dx_n$, and $ds$ is the surface element. Furthermore, if $u$ and $v$ are solutions of equations (1.2) and (1.60) respectively, by (1.62) we have

$$\text{Re} \int_S v(y)\mathbf{n}(y)u(y)ds_y = 0, \tag{1.64}$$

known as Green's formula.

To obtain Cauchy's integral formula we need to prove the following.

**Lemma.** *Let $S_\varepsilon$ be a spherical surface in $\Omega$ with radius $\varepsilon$ and with center at the point $x(x_0, x_1, \ldots, x_n)$. Then for any $x \in \Omega$ the equality*

$$M \equiv \lim_{\varepsilon \to 0} \text{Re} \int_{S_\varepsilon} q^A(|x-y|)\mathbf{n}(y)u(y)ds_y = \omega_{n+1} u_A(x) e_A^2, \tag{1.65}$$

*is valid where*

$$q^A(r) = e_A[g_{n+1}(r)\partial_y] + \widetilde{he}_A g_{n+1}(r), \quad r = |x-y|, \tag{1.66}$$

*are the fundamental solutions of (1.60) as functions of $y$, and $\omega_{n+1}$ is the surface area of the unit sphere $S_1$ in $R^{n+1}$.*

*Proof.* Because $ds_y = \varepsilon^n ds_1$, $n = (y-x)/\varepsilon$ on $S_\varepsilon$, and $u \in C^1(\Omega)$, by (1.56) one can easily get

$$M = \lim_{\varepsilon \to 0} \text{Re} \int_{S_1} e_A u(x + \varepsilon y) ds_1 = \omega_{n+1} e_A^2 u_A(x).$$

**Generalized Cauchy integral formula.** Let the domain $\Omega^+ \equiv \Omega$ and $\Omega^-$ be the complement of $\Omega^+ \cup S$ in the space $R^{n+1}$. Let $u(x)$ be a regular solution of equation (1.2) in $\Omega^+$, $u(x) \in C^1(\Omega^+) \cap C(\overline{\Omega^+})$. Consider the expression

$$K_s(u, h, x) \equiv \frac{1}{\omega_{n+1}} \int_S [\partial_y g_{n+1}(r)\mathbf{n}(y)u(y) + g_{n+1}(r)\overline{n}(y)\widetilde{u}(y)h] ds_y, \tag{1.67}$$

where $r = |x - y|$, for which we will prove

$$K_s(u, h, x) = \begin{cases} u(x), & \text{for } x \in \Omega^+, \\ 0, & \\ & \text{for } x \in \Omega^-. \end{cases} \tag{1.68}$$

## II. Multidimensional Cases

*Proof.* Let $\Omega_\varepsilon$ be the domain bounded by $S$ and $S_\varepsilon$. For this domain formula (1.64) can be used for $u(x)$ and $v = q^A(r)$ defined by (1.66). Then considering $\varepsilon \to 0$ by (1.65) one can get

$$\frac{1}{\omega_{n+1}} e_A^2 \operatorname{Re} \int_S \left[(e_A \partial_y g_{n+1}(r) + \overline{\widetilde{h} e}_A g_{n+1}(r)\right]\mathbf{n}(y)u(y)ds_y =$$

$$= \begin{cases} u_A(x), \\ x \in \Omega^+, \\ 0, \\ x \in \Omega^-. \end{cases}$$

We can obtain

$$u_A = e_A^2 \operatorname{Re}[ue_A],$$

$$u = \sum u_A e_A = \sum_A e_A^2 \operatorname{Re}[ue_A]e_A,$$

$$\operatorname{Re}[\overline{\widetilde{h}}_A \mathbf{n} u] = \operatorname{Re}[e_A \overline{\widetilde{\mathbf{n}} \widetilde{u}} h].$$

From these last equalities (1.68) follows easily. Using the equality $\overline{\partial}_x g_{n+1} = -\overline{\partial}_y g_{n+1}$, one check that (1.67) satisfies equation (1.2).

*Borel–Pompeiu representation.* Let $u$ be any function of the class $C^1(\Omega^+) \cap C(\overline{\Omega}^+)$ with values in $R_{(n)}$. Then using (1.62), (1.67) one can obtain the representation

$$K_S(u, h, x) - Tu = \begin{cases} u(x), \\ u \in \Omega^+, \\ 0, \\ x \in \Omega^-, \end{cases} \quad (1.69)$$

and the operator $T$ is defined as

$$Tu = \frac{1}{\omega_{n+1}} \int_{\Omega^+} \left[(\partial_y g_{n+1}(r))Lu + g_{n+1}(r)(\widetilde{L}u)h\right]d\Omega,$$

where $\widetilde{L}u = \partial \widetilde{u} + u\overline{h}$. $T$ can be considered as a multidimensional analogue of the operator considered in the two-dimensional case in [Ve] and for the quaternionic functions in [GS].

**Generalized Cauchy-type integral.** Let a continuous function $q(x)$ with values in $R_{(n)}$ be given only on $S$. Consider the integral

$$u(x) = K_S(q, h, x), \quad (1.70)$$

defined as in (1.67). This satisfies equation (1.2) everywhere in $R^{n+1} \setminus S$. By (1.53) it is clear that the second addend of the right-hand side of (1.70) with (1.67) exists for $x \in S$ as the usual integral, but the first addend is singular.

**Theorem.** *Let* $q(x) \in C^\alpha(S)$, $0 < \alpha \leq 1$; *i.e., each of its components* $q_A(x)$ *is a Hölder-continuous function*

$$|q_A(x) - q_A(y)| \leq M|x-y|^\alpha \quad \text{for any} \quad x, y \in S.$$

*Then the integral in* (1.70) *exists in the sense of a Cauchy principal value.*

The method of calculation of the Cauchy principal value is exactly the same as that for the Cauchy-type integral on the line in the two-dimensional case. Let $x \in S$ and $\Gamma_\varepsilon$ be that part of a spherical surface with radius $\varepsilon$ and center $x$ that is outside of $S$. Let $S_\varepsilon$ denote the part of the surface $S$ that is outside this spherical surface. Because we have

$$n = \frac{y-x}{|y-x|}, \quad \frac{(\bar{y}-\bar{x})\mathbf{n}(y)}{|y-x|^{n+1}} = \frac{1}{\varepsilon^n}, \quad ds_y = \varepsilon^n ds_1,$$

on $\Gamma_\varepsilon$, it is easy to obtain

$$\lim_{\varepsilon \to 0} \int_{\Gamma_\varepsilon} \frac{(\bar{y}-\bar{x})\mathbf{n}(y)}{|y-x|^{n+1}} ds_y = \frac{1}{2}\omega_{n+1}, \quad x \in S.$$

The Cauchy principal value of the following integral is by definition

$$P \equiv \int_S \frac{(\bar{y}-\bar{x})\mathbf{n}(y)}{|y-x|^{n+1}} ds_y = \lim_{\varepsilon \to 0} \int_{S_\varepsilon} \frac{(\bar{y}-\bar{x})\mathbf{n}(y)}{|y-x|^{n+1}} ds_y.$$

Considering (1.68) for the regular function $u(x) = 1$ and for $h = 0$ we have

$$\int_{S_\varepsilon \cup \Gamma_\varepsilon} \frac{(\bar{y}-\bar{x})\mathbf{n}(y)}{|y-x|^{n+1}} ds_y = \omega_{n+1}.$$

Thus we get

$$P = \frac{1}{2}\omega_{n+1}. \tag{1.71}$$

First the theorem will be proved for the integral (1.70) with $h = 0$, i.e., for the integral

$$\Phi(x) = \frac{1}{\omega_{n+1}} \int_S \left(\partial_y \frac{1}{r^{n-1}}\right) \mathbf{n}(y) q(y) \, dS_y, \quad r = |x-y|, \tag{1.72}$$

98    II. Multidimensional Cases

which is a Cauchy-type integral corresponding to regular functions. Now it is easy to prove that the Cauchy principal value of $\Phi(x)$ when $x \in S$ exists and by (1.71) we obtain

$$\Phi(x) = \frac{1}{\omega_{n+1}} \int_S \left(\partial_y \frac{1}{r^{n-1}}\right) \mathbf{n}(y)[q(y) - q(x)] \, dS_y + \frac{1}{2} q(x), \quad x \in S. \quad (1.73)$$

Because $q(x) \in C^\alpha(S)$, this integral exists in the usual sense.

Now let $x \in \Omega^+$ or $x \in \Omega^-$ and define the limit of (1.72) as $x \to x_0 \in S$. If we represent $\Phi(x)$ in the form

$$\Phi(x) = \frac{1}{\omega_{n+1}} \int_S \left(\partial_y \frac{1}{r^{n-1}}\right) \mathbf{n}(y)[q(y) - q(x_0)] \, dS_y +$$
$$+ \frac{1}{\omega_{n+1}} \int_S \left(\partial_y \frac{1}{r^{n-1}}\right) \mathbf{n}(y) \, dS_y, \ q(x_0)$$

and take into consideration the Cauchy integral formula (1.68) for $h = 0$, then because any constant is a regular function, we obtain

$$\Phi(x) = \frac{1}{\omega_{n+1}} \int_S \left(\partial_y \frac{1}{r^{n-1}}\right) \mathbf{n}(y)[q(y) - q(x_0)] \, dS_y + \begin{cases} q(x_0), & \text{if } x \in \Omega^+, \\ 0, & \text{if } x \in \Omega^-. \end{cases}$$

Then it is possible to consider $x \to x_0$, which by (1.71) can be written in the form

$$\Phi^+(x_0) = \frac{1}{2} q(x_0) + \Phi(x_0),$$
$$\Phi^-(x_0) = -\frac{1}{2} q(x_0) + \Phi(x_0). \quad (1.74)$$

These are the well-known Plemelj–Sokhotzki formulae for Cauchy-type integrals with Cauchy kernels corresponding to regular functions with values in $R^n$.

It is not difficult to obtain analogous formulae for the generalized Cauchy-type integral (1.70). Indeed, representing the first addend in the integrand of (1.70) with (1.67) as

$$\left[\partial_y \left(g_{n+1}(r) - \frac{1}{r^{n-1}}\right) + \partial_y \left(\frac{1}{r^{n-1}}\right)\right] \mathbf{n} q(y),$$

and then using (1.56), (1.74), one can obtain for (1.70) the Plemelj–Sokhotzki formulae as in (1.74). From them it follows that

# 1. BVP for ... Hobson Formula in Clifford Analysis

$$u^+(x) - u^-(x) = q(x),$$
$$u^+(x) + u^-(x) = 2u(x), \quad x \in S. \tag{1.75}$$

If we apply the Plemelj–Sokhotzki formulae to the Cauchy integral formula (1.68) one can easily obtain

$$K_S(u^+, h, x) = \frac{1}{2} u^+(x), \quad x \in S. \tag{1.76}$$

Let $\Omega^+ \in R^{n+1}$ be a domain with the boundary $S$ which is assumed to be a piecewise Lyapunov manifold. Let $\Omega^-$ denote the complement of $\Omega^+ \cup S$ in the whole space $R^{n+1}$.

**Problem** (Hilbert). Find a piecewise regular function $u(x)$ with values in $R_{(n)}$ that vanishes at infinity by the conditions

$$\bar{\partial} u = 0, \quad x \in \Omega^+, \Omega^-, \tag{1.77}$$
$$u^+(x) = u^-(x) G(x) + g(x), \quad x \in S, \tag{1.78}$$

where $G(x) \neq 0$, $g(x)$ are given Hölder-continuous functions with values in $R_{(n)}$ ($n \geq 1$).

In the case $n = 1$ this problem has been solved explicitly for wide classes of $G$, $g$ and $S$. For example [Mu1], when $S$ consists of a finite number of nonintersecting closed or open smooth contours, the functions $G$, $g$ satisfy the Hölder condition on $S$ except at a finite number of points of discontinuity of the first kind. (This class of functions is called the Muskhelishvili class.)

In the case $n > 1$ the solution of this problem can be represented in quadratures only in a few classes of $G$. The main reason for this fact is that the product of the two solutions of equation (1.77) is also a solution only in two-dimensional space $R_{(1)}$, but in $R_{(n)}$ ($n > 1$), the product is not a solution since we have the rule [Ob1]

$$\bar{\partial}(uv) = (\bar{\partial} u)v + \bar{u}(\bar{\partial} v) + 2 \operatorname{Re}[u\bar{\partial}]v - 2\bar{u} \frac{\partial v}{\partial x_0}.$$

If $G$ is constant, it is clear that $\bar{\partial} u G = (\bar{\partial} u) G$. Therefore, if $u$ is regular function, then $uG$ is regular too. Thus in (1.78) $G$ is supposed constant. Because we need $G^{-1}$, i.e., the inverse element of $G$, we suppose that $G$ is a vectorial element of $R_{(n)}$.

**Solution.** Let $\chi(x)$ be defined as a canonical function

$$\chi(x) = \begin{cases} G, \\ x \in \Omega^+, \\ e_0, \\ x \in \Omega^-, \end{cases}$$

i.e., $\chi^+(x) = \chi^-(x)G$. Then condition (1.78) can be rewritten in the form

$$u^+(x)[\chi^+(x)]^{-1} = u^-(x)[\chi^-(x)]^{-1} + g(x)[\chi^+(x)]^{-1}, \quad x \in S.$$

It is clear that $u(x)[G]^{-1}$, i.e., $u(x)[\chi(x)]^{-1}$, is also a solution of equation (1.77). By virtue of (1.74), the unique solution of (1.78) can be represented as

$$u(x) = \frac{1}{\omega_{n+1}} \int_S \frac{\bar{x} - \bar{y}}{|x - y|^{n+1}} \mathbf{n}(y) g(y) [\chi^+(y)]^{-1} dS_y \, \chi(x). \qquad (1.79)$$

The Hilbert problem for generalized regular equations must be modified in order to be solved explicitly. The point is that if $u(x)$ is regular and $G$ is constant, then $u(x)G$ is also regular. But if $u(x)$ is generalized regular, then $u(x)G$ is not generalized regular.

First note that if $g$ and $h$ are vectorial elements of $R_{(n)}$, then $h' = \overline{G}hG^{-1}$ is also vectorial and $|h'| = |h|$, which can be checked directly.

**Problem** (Hilbert). Let $u(x)$ with values in $R_{(n)}$ be the solution of the equations

$$\begin{aligned} \overline{\partial} u(x) + \tilde{u}h &= 0, \quad x \in \Omega^+, \\ \overline{\partial} u(x) + \tilde{u}h' &= 0, \quad x \in \Omega^-. \end{aligned} \qquad (1.80)$$

Define the solution of these equations that vanishes at infinity by the condition

$$u^+(x) = u^-(x)G + g(x), \quad x \in S, \qquad (1.81)$$

where $G \neq 0$ is vectorial constant and $g(x)$ is a Hölder-continuous function with values in $R_{(n)}$.

**Solution.** The function

$$v(x) = \begin{cases} u(x), & x \in \Omega^+, \\ u(x)G, & x \in \Omega^-, \end{cases} \qquad (1.82)$$

satisfies the first equation of (1.80) in $\Omega^+$ as well as in $\Omega^-$. By (1.81) one has

$$v^+ - v^- = g(x), \quad x \in S. \qquad (1.83)$$

By (1.75), (1.81) $v(x)$ and then $u(x)$ by (1.82) are defined uniquely in quadratures.

For generalized regular functions the Riemann–Hilbert and Compound BVP can be solved in analogous cases [Ob1].

**Poincaré–Bertrand transformation formula.** Let $q(y, z)$ with values in $R_{(n)}$ be a Hölder-continuous function of two points $y, z \in S$, where $S$ is a closed smooth surface. Consider two repeated integrals for $x \in S \subset R^{n+1}$

$$A(x) \equiv \int_S K_1(x, y) ds_y \int_S K_2(y, z) q(y, z) ds_z,$$
$$B(x) \equiv \int_S ds_z \int_S K_1(x, y) K_2(y, z) q(y, z) ds_y. \quad (1.84)$$

If the kernels $K_1, K_2$ are functions such that at least one integral exists on $S$ in the ordinary sense and the other exists in the sense of Cauchy's principal value, then inversion of the order of integration is legitimate, i.e.,

$$A(x) = B(x), \quad x \in S. \quad (1.85)$$

However if both integrals exist on $S$ in the sense of Cauchy's principal value the inversion is invalid. First of all the Poincaré–Bertrand transformation formula is obtained for the Cauchy-type integral (1.72), i.e.,

$$K_1(x, y) = K_2(x, y) = [\partial_y \phi(x - y)] \mathbf{n}(y), \quad (1.86)$$

where

$$\phi(x) = |x|^{1-n}.$$

**Theorem.** *For the repeated integrals* (1.84) *with the kernels* (1.86), *the following formula holds*

$$A(x) = \frac{1}{4} \omega_{n+1}^2 q(x, x) + B(x), \quad x \in S, \quad (1.87)$$

*where $\omega_{n+1}$ is defined by* (28).

*Proof.* Let $x \in \Omega^+$ or $\Omega^-$. Then the inversion (1.85) is legitimate since one of the singularities of the integrand, namely $x = y$, has been removed. Thus we have the equality (1.85). Then by (1.74), (1.84), (1.86) one can obtain for $x \to x_0 \in S$

$$A^+(x_0) + A^-(x_0) = 2 \int_S [\partial_y \phi(x_0 - y)] \mathbf{n}(y) ds_y \int_S [\partial_z \phi(y - z)] \mathbf{n}(z) q(y, z) ds_z. \quad (1.88)$$

102    II. Multidimensional Cases

Now consider $\overline{B(x)}$, which by (7), (1.84), (1.86) can be represented as

$$\overline{B(x)} = \int_S ds_z \int \overline{q(y,z)\,\mathbf{n}(z)[\overline{\partial}_z\phi(y-z)]\overline{\mathbf{n}}(y)[\overline{\partial}_y\phi(x-y)]} ds_y =$$
$$= \int_S \overline{q(z,z)\,\mathbf{n}(z)[\overline{\partial}_z\phi(x-z)]} ds_z \int_S \overline{\mathbf{n}(y)[\overline{\partial}_y\phi(x-y)]} ds_y +$$
$$+ \int_S \overline{q(z,z)\,\mathbf{n}(z)} ds_z \int_S \overline{[\overline{\partial}\phi(y-z) - \overline{\partial}_z\phi(x-z)]\overline{\mathbf{n}}(y)[\overline{\partial}_y\phi(x-y)]} ds_y +$$
$$+ \int_S ds_z \int_S \overline{[q(y,z) - q(z,z)]\mathbf{n}(z)[\overline{\partial}_z\phi(y-z)]\overline{\mathbf{n}}(y)[\overline{\partial}_y\phi(x-y)]} ds_y.$$

Then, taking into consideration (1.67), (1.76) for $h = 0$ and $u = 1$, we have

$$\int_S \overline{\mathbf{n}(y)[\overline{\partial}_y\phi(x-y)]} ds_y = \begin{cases} \omega_{n+1}, & x \in \Omega^+, \\ \frac{1}{2}\omega_{n+1}, & x \in S, \\ 0, & \\ & x \in \Omega^-. \end{cases}$$

Then by (1.74) for $\overline{B(x)}$ it follows that

$$\overline{B^+(x_0)} + \overline{B^-(x_0)} = \frac{1}{2}\omega_{n+1}^2 \overline{q(x_0,x_0)} +$$
$$+ 2\int_S ds_z \int_S \overline{q(y,z)\,\mathbf{n}(z)[\overline{\partial}_z\phi(y-z)]\overline{\mathbf{n}}(y)[\overline{\partial}_y\phi(x_0-y)]} ds_y. \qquad (1.89)$$

Therefore we have by the equality $A(x) = B(x)$ for $x \in \Omega^+$ or $\Omega^-$ when $x \to x_0 \in S$

$$A^+(x_0) + A^-(x_0) = B^+(x_0) + B^-(x_0).$$

Then, using (1.88), (1.89), one can obtain the formula (1.87).

Note that this remarkable formula for Cauchy-type singular integrals on the line in the case $n = 1$ is well known [Mu1] and has essential applications. The method of proof of the above theorem for $n > 1$ is almost the same as for $n = 1$.

Now, to prove formula (1.87) where $K_1$ and $K_2$ are defined by the integrand of (1.70), it is sufficient to consider the expression

$$K_1(x,y) = K_2(x,y) = [\partial_y g_{n+1}(|x-y|)]\mathbf{n}(y).$$

In fact, consider the representation

$$[\partial_y g_{n+1}(|x-y|)\mathbf{n}(y)[\partial_z g_{n+1}(|y-z|)]\mathbf{n}(z) =$$
$$= [\partial_y g_{n+1}(|x-y|) - \partial_y \phi(x-y)]\mathbf{n}(y)[\partial_z g_{n+1}(|y-z|)]\mathbf{n}(z) +$$
$$+ [\partial_y \phi(x-y)]\mathbf{n}(y)[\partial_z g_{n+1}(|y-z|) - \partial_z \phi(y-z)]\mathbf{n}(z) +$$
$$+ [\partial_y \phi(x-y)]\mathbf{n}(y)[\partial_z \phi(y-z)]\mathbf{n}(z),$$

where $\phi(x)$ is defined as in (1.86). Taking into consideration (1.46), formula (1.87) applies also to $K_1$ and $K_2$ defined by the generalized Cauchy-type integral (1.70).

**Inversion of the simplest singular integral equation.** Let $S \subset R^{n+1}$ be a closed Liapunov surface, $x_0 \in S$, and $p(x_0)$ be a given Hölder-continuous function. Define $q(x)$ to be a Hölder-continuous function on $S$ by the equation

$$\frac{2}{\omega_{n+1}} \int_S [(\partial_y g_{n+1}(|x_0 - y|))\mathbf{n}(y)q(y) + g_{n+1}(|x_0 - y|)\overline{\mathbf{n}}(y)\overline{q}(y)h] ds_y = p(x_0), \tag{1.90}$$

where the functions $p(x)$, $q(x)$ take values in $R_{(n)}$.

**Solution.** Consider the piecewise $h$-regular function that vanishes at infinity defined by (1.70)

$$u(x) = K_S(q, h, x), \tag{1.91}$$

where $x \in \Omega^+$ or $x \in \Omega^-$. Using (1.40) equation (1.90) can be written in the form

$$u^+(x_0) + u^-(x_0) = p(x_0), \quad x_0 \in S. \tag{1.92}$$

Now consider a second piecewise $h$-regular function with values in $R_{(n)}$ defined as

$$v(x) = \begin{cases} u(x), & x \in \Omega^+, \\ -u(x), & x \in \Omega^-. \end{cases} \tag{1.93}$$

Then (1.92) may be rewritten as

$$v^+(x_0) - v^-(x_0) = p(x_0),$$

so that by (1.75) for $x \in \Omega^+$ or $x \in \Omega^-$,

$$v(x) = K_S(p, h, x). \tag{1.94}$$

On the other hand, by (1.75), (1.91), (1.93) one can get

$$q(x_0) = u^+(x_0) - u^-(x_0) = v^+(x_0) + v^-(x_0),$$

and finally, taking into consideration (1.75), (1.94) we have

$$q(x_0) = 2K_S(p, h, x_0), \quad x_0 \in S. \tag{1.95}$$

It is obvious from this that $p(x_0)$ can be defined by (1.90). Hence (1.95) is the solution of equation (1.90). Thus each of equations (1.90) and (1.95) follows from the other; i.e., these relations are reciprocal.

## 2 BVP for Pluriregular, Generalized Pluriregular, and Polyharmonic Functions and the Poly-Helmholtz Equation

Let $u(x)$ be a pluriregular function in $R_{(n)}$, i.e., a solution of the pluriregular equation

$$\overline{\partial}^m u = 0, \quad m \geq 2. \tag{2.1}$$

Then $u(x)$ is also a solution of the polyharmonic equation

$$\Delta^m u = 0, \tag{2.2}$$

where $\Delta$ is the Laplace operator with respect to $x_0, x_1, \ldots, x_n$.

First consider BVP for (2.2), which help when solving BVP for (2.1). The solution of (2.2) can be represented as

$$u(x) = \sum_{0}^{m-1} x_n^k u_k(x), \tag{2.3}$$

or

$$u(x) = \sum_{0}^{m-1} (r^2 - 1)^k u_k(x), \quad r^2 = |x|^2, \tag{2.4}$$

where $u_k(x)$ are harmonic functions.

Let $D^+$ be the half space $x_n > 0$, $S^+$ be the circular domain $x_0^2 + x_1^2 \leq a^2$, and $S^-$ the domain $x_0^2 + x_1^2 > a^2$. $D$ is the ball $|x| \leq 1$.

**Problem 1.** Find the solution of (2.2) in $D^+$ that vanishes at infinity by the conditions on $x_n = 0$

$$\frac{\partial^k u}{\partial x_n^k} = f_k(x_0, \ldots, x_{n-1}), \quad k = 0, 1, \ldots, m-1. \tag{2.5}$$

**Solution.** By these conditions and using (2.3) we obtain the boundary conditions for $u_k$

$$\begin{aligned}
u_0(x_0, \ldots, x_{n-1}, 0) &= f_0, \\
u_1(x_0, \ldots, x_{n-1}, 0) + \frac{\partial u_0}{\partial x_n} &= f_1, \\
2u_2 + 2\frac{\partial u_1}{\partial x_n} + \frac{\partial^2 u_0}{\partial x_n^2} &= f_2,
\end{aligned} \tag{2.6}$$

and so on. Because the left sides in (2.6) are boundary conditions of harmonic functions, by (1.13) one can define them all. For instance, if $m = 2$, we obtain

$$u(x) = \frac{2(n+1)x_n^3}{\omega_{n+1}} \int_{R^n} \frac{f_0(\xi)\,d\xi}{r^{n+3}} + \frac{2x_n^2}{\omega_{n+1}} \int_{R^n} \frac{f_1(\xi)\,d\xi}{r^{n+1}}. \tag{2.7}$$

For any $m \geq 2$,

$$u(x) = \sum_{k=0}^{m-1} \frac{2}{\omega_{n+1}k!} x_n^k \sum_{l=0}^{k} (-1)^l C_k^l \frac{d^l}{dx_n^l} x_n \int_{R^n} f_{k-l}(\xi) \frac{d\xi}{r^{n+1}}, \tag{2.8}$$

where $r^2 = (x_0 - \xi_0)^2 + \cdots + (x_{n-1} - \xi_{n-1})^2 + x_n^2$. These representations by FIT are obtained in [Ob1] too.

For biharmonic functions, consider problems like the Hobson problem for harmonic functions.

**Problem 2.** Find the biharmonic function in the half space $x_2 > 0$ that vanishes at infinity by the conditions

$$\begin{aligned}
u(x_0, x_1, 0) &= f_0(x_0, x_1), \quad (x_0, x_1) \in R^2, \\
\Delta u|_{x_2=0} &= f_1(x_0, x_1), \quad x_0^2 + x_1^2 > a^2,
\end{aligned} \tag{2.9}$$

$$\frac{\partial u}{\partial x_2}\bigg|_{x_2=0} = f_2(x_0, x_1), \quad x_0^2 + x_1^2 < a^2. \tag{2.10}$$

**Solution.** Because

$$u(x) = x_2 u_1 + u_0, \qquad (2.11)$$

for harmonic functions $u_1, u_2$, by force of (2.9), (2.10) we have the boundary conditions

$$u_0(x_0, x_1, 0) = f_0(x_0, x_1), \quad (x_0, x_1) \in R^2,$$

$$\Delta u|_{x_2=0} = 2\frac{\partial u_1}{\partial x_2} = f_1(x_0, x_1), \quad x_0^2 + x_1^2 > a^2,$$

$$\frac{\partial u}{\partial x_2}\bigg|_{x_2=0} = u_1 + \frac{\partial u_0}{\partial x_2} = f_2(x_0, x_1), \quad x_0^2 + x_1^2 < a^2.$$

By force of (1.13), $u_0$ is represented in quadratures. Then for $u_1$ we have conditions (1.38), (1.39). Thus the solution is represented using (1.41).

**Problem 3.** Find in the space with the crack along $x_0^2 + x_1^2 \leq a^2$ the biharmonic function that vanishes at infinity by the conditions

$$u^{\pm}(x_0, x_1, 0) = f^{\pm}(x_0, x_1), \quad x_0^2 + x_1^2 \leq a^2,$$

$$\frac{\partial u^{\pm}}{\partial x_2}\bigg|_{x_2=0} = \varphi^{\pm}(x_0, x_1).$$

**Solution.** For harmonic functions $u_0, u_1$ we have boundary conditions (1.35), the solutions of which are represented using the Hobson formula (1.33).

Problems for polyharmonic functions in the case $m > 2$ with the boundary conditions

$$\frac{\partial^k u}{\partial x_2^k}\bigg|_{x_2=0}^{\pm} = \varphi_k^{\pm}(x_0, x_1), \quad k = 0, \ldots, m-1,$$

by force of (2.3), can be reduced to the problems for harmonic functions with conditions like (1.35); i.e., the solution is also represented with the help of the Hobson formula.

For the ball $|x| \leq 1$, formula (2.4) is used.

**Problem 4.** Find the polyharmonic function $u(x)$ in the ball $|x| < 1$ with the boundary conditions

$$\frac{\partial^k u}{\partial r^k}\bigg|_{r=1} = f_k(x), \quad k = 0, 1, \ldots, m-1. \qquad (2.12)$$

**Solution.** By force of (2.4) for harmonic functions $u_k(x)$, $k = 0, \ldots, m-1$, we have the conditions

$$u_0(x) = f_0(x), \quad |x| = 1,$$

$$\frac{\partial u}{\partial r} = 2u_1 + \frac{\partial u_0}{\partial r} = f_1, \quad \text{for} \quad r = 1, \qquad (2.13)$$

$$\frac{\partial^2 u}{\partial r^2} = 8u_2 + 2u_1 + 4\frac{\partial u_1}{\partial r} + \frac{\partial^2 u_0}{\partial r^2} = f_2,$$

## 2. BVP for ... and the Poly-Helmholtz Equation

and so on. First we prove that if $\varphi$ is a harmonic function, then $r\frac{\partial \varphi}{\partial r}$ is harmonic. We have that

$$r\frac{\partial \varphi}{\partial r} = \sum_0^n \frac{\partial \varphi}{\partial x_k} x_k,$$

and

$$\Delta\left(r\frac{\partial \varphi}{\partial r}\right) = 2\sum_0^n \frac{\partial^2 \varphi}{\partial x_k^2} = 0.$$

By the inductive method, one can prove that if $r^{k-1}\frac{\partial^{k-1}\varphi}{\partial r^{k-1}}$ is harmonic, then $r^k\frac{\partial^k \varphi}{\partial r^k}$ is harmonic too:

$$r^k\frac{\partial^k \varphi}{\partial r^k} = r\frac{\partial}{\partial r}\left(r^{k-1}\frac{\partial^{k-1}\varphi}{\partial r^k}\right) - (k-1)r^{k-1}\frac{\partial^{k-1}\varphi}{\partial r^{k-1}}.$$

For this reason (2.13) are boundary conditions of harmonic functions

$$u_0(x), \quad 2u_1 + r\frac{\partial u_0}{\partial r}, \quad 8u_2 + 2u_1 + 4r\frac{\partial u_1}{\partial r} + r^2\frac{\partial^2 u_0}{\partial r^2},$$

and so on. Thus these harmonic functions are defined by (1.18), $u_0, u_1, u_2, \ldots$ are gradually defined, and correspondingly by (2.4), $u$ is represented in quadratures. For instance, for biharmonic functions we have

$$u(x) = \frac{1}{2}(r^2 - 1)Pf_1 - \frac{1}{2}(r^2 - 1)r\frac{\partial}{\partial r}Pf_0 + Pf_0,$$

where

$$Pf \equiv \frac{1 - r^2}{\omega_n} \int_{|\xi|=1} \frac{f(\xi)\, d\xi}{|\xi - x|^{n+1}}, \quad x \in R^{n+1}. \tag{2.14}$$

Now consider equation (2.1) for $m = 2$ and $n = 2$, $u = u_0 e_0 - u_1 e_1 - u_2 e_2 - u_{12} e_1 e_2$.

**Problem 5.** Find the solution of (2.1) in the half space $x_2 > 0$ that vanishes at infinity by the conditions

$$u_0(x_0, x_1, 0) = f_0(x_0, x_1), \quad \left.\frac{\partial u_2}{\partial x_2}\right|_{x_2=0} = f_1(x_0, x_1),$$

$$u_1(x_0, x_1, 0) = \varphi_0(x_0, x_1), \quad \left.\frac{\partial u_{12}}{\partial x_2}\right|_{x_2=0} = \varphi_1(x_0, x_1). \tag{2.15}$$

## II. Multidimensional Cases

**Solution.** Equation (2.1) can be written as

$$\overline{\partial} u = F, \quad \overline{\partial} F = 0, \quad F = F_0 e_0 - F_1 e_1 - F_2 e_2 - F_{12} e_1 e_2.$$

It is clear that

$$F_1 = \frac{\partial u_0}{\partial x_0} + \frac{\partial u_1}{\partial x_1} + \frac{\partial u_2}{\partial x_2}, \quad F_2 = \frac{\partial u_{12}}{\partial x_2} - \frac{\partial u_0}{\partial x_1} + \frac{\partial u_1}{\partial x_0}. \tag{2.16}$$

Thus by the given conditions, $F_1$, $F_2$ are given for $x_2 = 0$. Then $F$, as the solution of $\overline{\partial} F = 0$, is defined in the half space $x_2 > 0$ and $u$ is defined from nonhomogeneous equation $\overline{\partial} u = F$ with the conditions $u_0 = f_0$, $u_1 = \varphi_0$ for $x_2 = 0$.

Thus for the equation $\overline{\partial}^2 u = 0$ four conditions are sufficient. In other words, all the above problems solved for domains with plane cracks or for balls can be solved for this equation.

Now consider the plurigeneralized regular equation of $m$-th order

$$P^m u = 0, \tag{2.17}$$

where

$$Pu = \overline{\partial} u + \tilde{u} h, \quad h = \sum_0^n h_k e_k.$$

For $m = 2$ we have the bigeneralized regular equation.

Because the solution of the equation $\overline{\partial} u + \tilde{u} h = 0$ with $h$ constant is also the solution of the Helmholtz equation, one can obtain from (2.17) that $u$ is also the solution of the poly-Helmholtz equation

$$(\Delta - |h|^2)^m u = 0. \tag{2.18}$$

For $m = 2$ it is called the bi-Helmholtz equation.

Note that if $u_k$ ($k = 0, 1, \ldots, m-1$) are the solutions of the equation $\Delta u_k - |h|^2 u_k = 0$, then the representation

$$u = \sum_{k=0}^{m-1} x_n^k u_k(x),$$

is the solution of equation (2.18).

Using this, BVP with the conditions

## 3. BVP for ... and Pluri-Beltrami Equations in Clifford Analysis

$$\frac{\partial^k u}{\partial x_n^k} = f_k(x_0, \ldots, x_{n-1}), \quad x_n = 0, \quad k = 0, 1, \ldots, m-1,$$

in the half space $x_n > 0$ can be reduced to the Dirichlet problem for $u_k$ ($k = 0, \ldots, m-1$) which are represented in quadratures by (1.30). If we have boundary conditions (Riquie)

$$\Delta^k u = f_k(x_0, \ldots, x_{n-1}), \quad k = 0, \ldots, m-1, \quad x_n = 0, \tag{2.19}$$

equation (2.18) is represented in the form

$$(\Delta - |h|^2)^{m-1} u = F, \tag{2.20}$$
$$\Delta F - |h|^2 F = 0. \tag{2.21}$$

By the conditions (2.19) one can define $F$ for $x_n = 0$. Thus $F$ is defined by (1.30). Then $u$ is defined from (2.20) gradually.

It is interesting to consider the equation

$$\Delta(\Delta - |h|^2) u = 0, \quad u(x), \quad x(x_0, \ldots, x_n), \tag{2.22}$$

which can be called the harmonic-Helmholtz equation.

**Dirichlet Problem.** Find the regular solution of (2.22) for $x_n > 0$ that vanishes at infinity by the conditions

$$u(x_0, \ldots, x_{n-1}, 0) = \varphi(x_0, \ldots, x_{n-1}), \quad \frac{\partial^2 u}{\partial x_n^2} = \psi(x_0, \ldots, x_{n-1}), \quad x_n = 0. \tag{2.23}$$

**Solution.** Let

$$\Delta u = F(x), \tag{2.24}$$
$$\Delta F - |h|^2 F = 0, \quad x_n > 0. \tag{2.25}$$

Then by force of (2.23), (2.24) we get

$$F(x_0, \ldots, x_{n-1}, 0) = \Delta\varphi(x_0, \ldots, x_{n-1}) + \psi(x_0, \ldots, x_{n-1}) \equiv f(x_0, \ldots, x_{n-1}).$$

Thus $F$ is defined as the solution of (2.25) with this condition by (1.30). From (2.24) with the condition $u(x_0, \ldots, x_{n-1}, 0) = \varphi(x_0, \ldots, x_{n-1})$ $u$ can be defined effectively.

So we see that all problems solved for the Helmholtz equation can be solved for equations (2.18), (2.22).

## 3  BVP for Beltrami, Generalized Beltrami, and Pluri-Beltrami Equations in Clifford Analysis

Let $u(x)$ belong to $R_{(n)}$ ($n \geq 2$) and consider the Beltrami equation

$$\overline{\partial}u + q\partial u = 0, \quad x = \sum_{0}^{n} x_k e_k, \tag{3.1}$$

the generalized Beltrami equation

$$\overline{\partial}u + q_1\partial u + q_2\overline{\partial}\overline{u} = 0, \tag{3.2}$$

and the pluri-Beltrami equation

$$(\overline{\partial} + q\partial)^m u = 0, \quad m \geq 2. \tag{3.3}$$

**3.1 Classification in multidimensional space.** First we consider the Beltrami equation (3.1) in the spaces $R_{(n)}$ ($n \geq 2$), where $q$ is a variable vectorial element in this space. The surface

$$S : \varphi(x_0, \ldots, x_n) = 0, \quad \frac{\partial \varphi(x_0, \ldots, x_n)}{\partial x_n} \neq 0 \tag{3.4}$$

is characteristic for (3.1) if one cannot define all the first derivatives of $u$ by the given solution

$$u(x_0, \ldots, x_n) = f(x_0, \ldots, x_n) \quad \text{on} \quad S. \tag{3.5}$$

With

$$\frac{\partial \varphi}{\partial x_k} + \frac{\partial \varphi}{\partial x_n}\frac{\partial x_n}{\partial x_k} = 0, \quad k = 0, 1, \ldots, n-1, \quad \tau_k \equiv -\frac{\partial x_n}{\partial x_k},$$

and with (3.5) one can derive

$$\frac{\partial u}{\partial x_k} = \frac{\partial u}{\partial x_n}\tau_k + G_k \quad \text{on} \quad S, \quad k = 0, 1, \ldots, n-1. \tag{3.6}$$

Then substituting (3.6) into (3.1) we have

# 3. BVP for ... and Pluri-Beltrami Equations in Clifford Analysis

$$(\tau + q\bar{\tau})\frac{\partial u}{\partial x_n} = F_1 \quad \text{on} \quad S, \tag{3.7}$$

where $\tau = \sum_{k=0}^{n} \tau_k e_k$, $\tau_n = 1$. From this equation we obtain

$$(\bar{\tau} + \tau\bar{q})(\tau + q\bar{\tau})\frac{\partial u}{\partial x_n} = F_2 \quad \text{on} \quad S. \tag{3.8}$$

Because $\tau$ and $q$ are vectorial elements of the spaces, it is easy to check directly that $\tau\bar{q}\tau$ contains only vectorial and bivectorial parts; thus

$$Q(\tau) = (\bar{\tau} + \tau\bar{q})(\tau + q\bar{\tau}), \tag{3.9}$$

is real and $\dfrac{\partial u}{\partial x_n}$ can be defined uniquely on $S$ from (3.8) if $Q(\tau) \neq 0$.

**Definition.** If $Q(\tau)$ is a definite form with respect to $\tau_0, \ldots, \tau_n$, then (3.1) is an elliptic equation. If $Q$ is an indefinite form, then (3.1) is hyperbolic. And if $Q$ is degenerate, then (3.1) is parabolic.

We consider (3.1), $\tau, q \in R_{(n)}$. By force of the above indicated property of $\tau\bar{q}\tau$ it is obvious that

$$\tau\bar{q}\tau + \bar{\tau}q\bar{\tau} = 2\,\text{Re}[\tau\bar{q}\tau] = 2q_0\left(\tau_0^2 - \sum_{k=1}^{n}\tau_k^2\right) + 4\tau_0\sum_{k=1}^{n}q_k\tau_k.$$

Thus, by (3.9), $Q(\tau)$ is defined as

$$Q(\tau) = \tau_0^2(1 + |q|^2 + 2q_0) + \sum_{k=1}^{n}\tau_k^2(1 + |q|^2 - 2q_0) + 4\tau_0\sum_{k=1}^{n}q_k\tau_k. \tag{3.10}$$

The discriminant of $Q$,

$$N(q) = \begin{vmatrix} 1+|q|^2+2q_0 & 2q_1 & \cdots & \cdots & 2q_n \\ 2q_1 & 1+|q|^2-2q_0 & 0 & \cdots & 0 \\ \vdots & \vdots & \vdots & \vdots & \vdots \\ 2q_n & 0 & \cdots & 0 & 1+|q|^2-2q_0 \end{vmatrix}, \tag{3.11}$$

can be calculated by mathematical induction to be

$$N(q) = (1+|q|^2 - 2q_0)^{n-1}[(1+|q|^2)^2 - 4q_0^2 - \cdots - 4q_n^2] =$$
$$= \left[(1-q_0)^2 + \sum_{k=1}^{n} q_k^2\right]^{n-1}[1-|q|^2]^2. \tag{3.12}$$

Thus $Q(\tau)$ is a positive definite form if $N(q) > 0$, i.e., if

$$|q| \neq 1, \tag{3.13}$$

and degenerate if $|q| = 1$. As in the preceding two-dimensional case, equation (3.1) with condition (3.13) in the space $R_{(n)}$ is only elliptic [Ob2].

Equation (3.1) will be considered only for the condition

$$|q| < 1, \tag{3.14}$$

since the case $|q| > 1$ can be reduced to the same equation with $|q| < 1$ by considering the corresponding equation for $\tilde{u}$,

$$\partial \tilde{u} + \frac{q}{|q|^2} \bar{\partial} \tilde{u} = 0.$$

Consider the space $R_{(n,n-1)}$. Assuming that

$$q\bar{q} = |q|^2 = \sum_{lk=0}^{n-1} q_k^2 - q_n^2 > 0,$$

one can show as above that (3.1) is hyperbolic if $|q| \neq 1$ and that it is sufficient to consider the condition (3.14).

To classify the generalized Beltrami equation (3.2), where $q_1$ and $q_2$ are vectorial variables of $R_{(n)}$, we use (3.6) and, as in (3.7), we obtain

$$(\tau + q_1 \bar{\tau}) \frac{\partial u}{\partial x_n} + q_2 \tau \frac{\partial \bar{u}}{\partial x_n} = F \quad \text{on} \quad S.$$

In order to define $\dfrac{\partial u_A}{\partial x_n}$ for every $A = (\alpha_1, \ldots, \alpha_k)$, from this equation and its conjugate we derive

$$(\tau + q_1 \bar{\tau}) \frac{\partial u}{\partial x_n} e_A(\bar{\tau} + \tau \bar{q}_1) + q_2 \tau \frac{\partial \bar{u}}{\partial x_n} e_A(\bar{\tau} + \tau \bar{q}_1) = F_1 \quad \text{on} \quad S,$$
$$q_2 \tau \bar{e}_A \frac{\partial u}{\partial x_n} \bar{\tau} q_2 + q_2 \tau \bar{e}_A \frac{\partial \bar{u}}{\partial x_n} (\bar{\tau} + \tau \bar{q}_1) = F_2 \quad \text{on} \quad S. \tag{3.15}$$

## 3. BVP for ... and Pluri-Beltrami Equations in Clifford Analysis

The real part of $\dfrac{\partial u}{\partial x_n} e_A$ is $\dfrac{\partial u_A}{\partial x_n} e_A^2$ and the real part of $\bar{e}_A \dfrac{\partial \bar{u}}{\partial x_n}$ is $\dfrac{\partial u_A}{\partial x_n} \bar{e}_A^2$. Because $\tau$ and $\tau^2$ are vectorial, the expressions $q_1 \bar\tau^2$ and $\tau^2 \bar q_1$ contain only vectorial and bivectorial parts; therefore,

$$q_1 \bar\tau^2 + \tau^2 \bar q_1 = 2\,\mathrm{Re}[q_1 \bar\tau^2].$$

We add the real parts of equations (3.15) if $\bar{e}_A e_A = -e_A^2$ and subtract them if $\bar{e}_A e_A = e_A^2$; in both cases we obtain

$$Q(\tau) \dfrac{\partial u_A}{\partial x_n} = F_3 \quad \text{on} \quad S, \quad A = (\alpha_1, \ldots, \alpha_k),$$

where

$$Q(\tau) = (\tau + q_1 \bar\tau)(\bar\tau + \tau \bar q_1) - q_2 \tau \bar\tau \bar q_2 \tag{3.16}$$

is called the characteristic polynomial for (3.2). If $Q(\tau)$ is a definite quadratic form with respect to $\tau_0, \ldots, \tau_n$, then (3.2) is elliptic. If $Q(\tau)$ is indefinite, the equation is hyperbolic. And if $Q(\tau)$ is degenerate, it is parabolic.

In the space $R_{(n)}$, $Q(\tau)$ can be represented as

$$Q(\tau) = \tau_0^2(1 + |q_1|^2 - |q_2|^2 + 2q_{10}) +$$
$$+ \sum_{k=1}^{n} \tau_k^2(1 + |q_1|^2 - |q_2|^2 - 2q_{10}) + 4\tau_0 \sum_{k=1}^{n} q_{1k}\tau_k.$$

The discriminant of $Q$ can be calculated as in (3.11) and we find that

$$N(q_1, q_2) = [1 + |q_1|^2 - |q_2|^2 - 2q_{10}]^{n-1}[(1 + |q_1|^2 - |q_2|^2)^2 - 4|q_1|^2].$$

It is easily seen that, if $q_l = \sum_{k=0}^{n} q_{lk} e_k$, then

$$1 + |q_1|^2 - |q_2|^2 - 2q_{10} \geq 1 + |q_1|^2 - |q_2|^2 - 2|q_1|$$
$$= (1 - |q_1| - |q_2|)(1 - |q_1| + |q_2|).$$

In the space $R_{(n,n-1)}$ we can obtain an analogous representation for $Q(\tau)$ and its discriminant. Then for $Q(\tau)$ definite, indefinite, or degenerate, conditions similar to (4.8), (4.11) can be obtained.

The classification of the above equations is remarkable because it is the same for determined and overdetermined systems. If $u$ is vectorial, then (3.2) consists of $1 +$

$n(n+1)/2$ equations with $n+1$ ($n > 1$) unknowns, while if $u$ is a general element of the space, then there are $2^n$ equations with $2^n$ unknowns. The classification of the overdetermined system is given by different rules; it seems me that the Clifford analysis method of classification is simpler.

Equation (3.1) with vectorial constant coefficients $q$ in the space $R_{(n)}$ is reduced to the equation [Ob2]

$$\bar{\partial} u = 0, \qquad (3.17)$$

by a uniquely defined linear transformation. Thus a solution of (3.1) is represented by a solution of (3.17).

**3.2 BVP.** Because we are interested in solving problems explicitly in quadratures, the above equations will be considered with constant coefficients.

First consider the Beltrami operator with the variable $x = \sum_{k=0}^{n} x_k e_k$,

$$Bu \equiv \bar{\partial}_x u + q \partial_x u, \quad q = \sum_{k=0}^{n} q_k e_k, \quad |q| \neq 1. \qquad (3.18)$$

**Theorem.** *The linear transformation*

$$y_k = \sum_{i=0}^{n} a_{ki} x_i, \quad k = 0, 1, \ldots, n,$$

*can be defined in such a way that for any real $P$, with $P^2 \neq 1$, the operator $B$ is reduced to the form*

$$Bu = \bar{\partial}_y u + P \partial_y u = (1+P) \frac{\partial u}{\partial y_0} + (1-P) \sum_{k=1}^{n} e_k \frac{\partial u}{\partial y_k}. \qquad (3.19)$$

*Proof.* In the previous section we saw that the case $|q| > 1$ can be reduced to $|q| < 1$; thus the proof will be given assuming that $|q| < 1$.

Because $\dfrac{\partial}{\partial x_j} = \sum_{k=0}^{n} a_{kj} \dfrac{\partial}{\partial y_k}$, to obtain equality (3.19) it is sufficient that

$$\sum_{j=0}^{n} e_j a_{oj} + \sum_{i=0}^{n} q_i e_i \left( a_{00} - \sum_{j=1}^{n} e_j a_{0j} \right) = 1 + P, \qquad (3.20)$$

$$\sum_{j=0}^{n} e_j a_{kj} + \sum_{i=0}^{n} q_i e_i \left( a_{k0} - \sum_{j=1}^{n} e_j a_{kj} \right) = (1-P) e_k, \quad k = 1, \ldots, n. \qquad (3.21)$$

Equations (3.20), (3.21) are a system for the unknown coefficients $a_{ki}$. For $a_{0j}$, $j = 0, \ldots, n$, we have

$$a_{00}(1 + q_0) + \sum_{j=1}^{n} q_j a_{0j} = 1 + P, \tag{3.22}$$

$$a_{00} q_j + (1 - q_0) a_{0j} = 0, \quad j = 1, 2, \ldots, n, \tag{3.23}$$

as well as the equations

$$q_j a_{0k} - q_k a_{0j} = 0,$$

which are not independent because they follow from (3.22), (3.23). Since the determinant of this system is

$$M(q) = \begin{vmatrix} 1 + q_0 & q_1 & \cdots & \cdots & q_n \\ q_1 & 1 - q_0 & 0 & \cdots & 0 \\ \vdots & \vdots & \vdots & \vdots & \vdots \\ q_n & 0 & \cdots & 0 & 1 - q_0 \end{vmatrix} = (1 - q_0)^{n-1}(1 - |q|^2) \neq 0,$$

(3.24)

the solution is defined uniquely. We have

$$a_{00} = \frac{(1 + P)(1 - q_0)}{1 - |q|^2}, \quad a_{0j} = -\frac{(1 + P) q_j}{1 - |q|^2}, \quad j = 1, \ldots, n. \tag{3.25}$$

For the other $a_{kj}$ ($j = 0, 1, \ldots, n$), with each fixed $k \geq 1$ we have a system with the same determinant (3.24) and with the right side of the $k$th equation equal to $1 - P$ and the right side of all other equations equal to 0. Thus all $a_{kj}$ are uniquely and explicitly defined.

We see that $P$ can be zero, in which case the above linear transformation reduces the operator (3.18) to the regular operator (3.19). Thus the solution of equation (3.18) can be represented as

$$u(x) = \varphi(y_0, y_1, \ldots, y_n), \tag{3.26}$$

where $\varphi(y)$ is a solution of the equation

$$\bar{\partial}_y \varphi = 0. \tag{3.27}$$

In particular, when $n = 1$, from (3.20), (3.21), and (3.25) we have the well-known representation

$$u(x) = \varphi(\zeta), \quad \zeta = \frac{z - q\bar{z}}{1 - |q|^2}, \quad |q| \neq 1, \tag{3.28}$$

where $\varphi(\zeta)$ is a holomorphic function of one complex variable $\zeta$.

Now consider the equation

$$\bar{\partial}_x u + q \partial_x u + \tilde{u} h = 0, \tag{3.29}$$

where $q$ and $h$ are vectorial constants of the space $R_{(n)}$. It is clear that, by the same linear transformation defined by (3.20), (3.21) with $P = 0$, this equation can be reduced to

$$\bar{\partial}_y u + \tilde{u} h = 0. \tag{3.30}$$

Thus a solution of (3.29) can be represented by (3.26), where $\varphi$ is a solution of (3.30). Moreover, all problems that were solved above for equations (3.27), (3.30), can be solved for equations (3.18), (3.29) with corresponding conditions.

The solution of (3.2) with constants $q_1, q_2$ in the two-dimensional case was reduced above to the solution of the Cauchy–Riemann system. In the multidimensional case it is not possible to reduce it to equation (3.27).

But consider the equation with vectorial constants $q_1, q_2 \in R_{(n)}$

$$\bar{\partial} u + q_1 \partial u + q_2 \overline{\partial \tilde{u}} = 0. \tag{3.31}$$

First, that equation can be reduced to an equation in terms of real $q_2$. If $q_2$ is represented as any vectorial element then

$$P = |P|\left(\frac{P_0 e_0}{|P|} + \cdots + \frac{P_n e_n}{|P|}\right) =$$

$$= |P|\left[\frac{\sqrt{P_0 + |P|}}{\sqrt{2|P|}} + \frac{P_1 e_1}{\sqrt{2|P|}\sqrt{P_0 + |P|}} + \cdots + \frac{P_n e_n}{\sqrt{2|P|}\sqrt{P_0 + |P|}}\right]^2.$$

If $P$ is represented in the spherical coordinates

$$P = |P|(e_0 \cos \alpha_0 + e_1 \sin \alpha_0 \cos \alpha_1 + e_2 \sin \alpha_0 \sin \alpha_1 \cos \alpha_2 + \cdots +$$
$$+ e_n \sin \alpha_0 \sin \alpha_1 \sin \alpha_2 \cdots \sin \alpha_{n-1}) =$$
$$= |P|\left[e_0 \cos \frac{\alpha_0}{2} + e_1 \sin \frac{\alpha_0}{2} \cos \alpha_1 + e_2 \sin \frac{\alpha_0}{2} \sin \alpha_1 \cos \alpha_2 + \cdots +\right.$$
$$\left. + e_n \sin \frac{\alpha_0}{2} \sin \alpha_1 \sin \alpha_2 \cdots \sin \alpha_{n-1}\right]^2 \equiv$$
$$\equiv |P|(a_0 e_0 + a_1 e_1 + \cdots + a_n e_n)^2.$$

## 3. BVP for ... and Pluri-Beltrami Equations in Clifford Analysis

Note that we have the De Moivre formula in a multidimensional space

$$\left(e_0 \cos \frac{\alpha_0}{m} + e_1 \sin \frac{\alpha_0}{m} \cos \alpha_1 + \cdots + e_n \sin \frac{\alpha_0}{m} \sin \alpha_1 \cdots \sin \alpha_{n-1}\right)^m =$$
$$= (e_0 \cos \alpha_0 + e_1 \sin \alpha_0 \cos \alpha_1 + \cdots + e_n \sin \alpha_0 \cdots \sin \alpha_{n-1}).$$

Then for the new function

$$w = (a_0 e_0 + a_1 e_1 + \cdots a_n e_n)^{-1} u,$$

equation (3.31) can be written in the form

$$\overline{\partial} w + q_1 \partial w + |q_2| \overline{\partial} \widetilde{w} = 0.$$

This equation can be reduced to a regular equation if $q_1$ is real. Thus, consider equation (3.31) with real coefficients. Then, for instance, in the case $n = 2$, by the transformation of the independent variables

$$y_0 = \left[(1 - q_1)^2 - q_2^2\right]^{1/2} x_0,$$
$$y_1 = \left[(1 + q_1)^2 - q_2^2\right]^{1/2} x_1,$$
$$y_2 = \left[(1 + q_1)^2 - q_2^2\right]^{1/2} x_2,$$

and by the transformation of the functions

$$v_0 = \left[(1 + q_2)^2 - q_1^2\right]^{1/2} u_0,$$
$$v_1 = \left[(1 - q_2)^2 - q_1^2\right]^{1/2} u_1,$$
$$v_2 = \left[(1 - q_2)^2 - q_1^2\right]^{1/2} u_2,$$
$$v_3 = \left[(1 + q_2)^2 - q_1^2\right]^{1/2} u_3,$$

equation (3.31) is transformed as

$$\overline{\partial}_y v = 0;$$

i.e., in this case one obtains a Moisil–Theodorescu system and the corresponding BVP can be solved in quadratures.

Now it is obvious that by the same linear transformation obtained above equation (3.3) can be reduced to equation (2.1) and the corresponding problems can be solved in quadratures too. Note that for equation (2.1) some general integral representations are obtained in [Be]; hence, the corresponding integral representations can be obtained for equation (3.3).

It is clear that $\overline{\partial}$ in $R_{(1)}$ is a Cauchy–Riemann operator. Thus all equations considered in the first chapter are two-dimensional cases of the equations considered in this chapter.

## 4 More Problems for Pluriregular and Plurigeneralized Regular Functions

Let $\Omega$ be a bounded domain in $R^{n+1}$ with a closed piecewise smooth Lyapunov surface $S$, $u(x): \Omega \to R_{(n)}$. Consider the pluriregular and plurigeneralized regular equations

$$\overline{\partial}^m u(x) = 0, \quad x(x_0, \ldots, x_n), \tag{4.1}$$

$$P^m u = 0, \quad Pu = \overline{\partial} u + \widetilde{u} h, \quad m \geq 1. \tag{4.2}$$

In the case $m = 1$, i.e., for regular and generalized regular functions, the Liouville theorem has been proved [Ob1]: *If $u(x)$ is a regular or generalized regular function in the space $R^{n+1}$ that vanishes at infinity, then $u(x)$ must be zero.* It is obvious that the Liouville theorem holds for equations (3.1), (3.3) too. For equations (4.1), (4.2) we can prove:

**Liouville theorem.** *Let $u(x)$ be the solution of (4.1) or (4.2) in the space $R^{n+1}$ that satisfies the conditions*

$$\lim_{|x| \to \infty} \overline{\partial}^k u(x) = 0, \quad k = 0, 1, \ldots, m-1. \tag{4.3}$$

*Then $u(x) \equiv 0$ for every $x \in R^{n+1}$.*

As in the case $m = 1$, the proof for $m > 1$ follows at once.

Let $\Omega_1$ and $\Omega_2$ be bounded domains in $R^{n+1}$ having no interior points in common with boundaries $S_1$ and $S_2$, respectively, which are Lyapunov surfaces and where $S_1 \cap S_2 = S_0$.

In the case $m = 1$ the following theorem is proved in [Ob1].

**Extension theorem.** *If $u(x)$ and $v(x)$ are regular or generalized regular in $\Omega_1$ and $\Omega_2$, respectively, and Hölder-continuous in the closed domains $\overline{\Omega}_1$ and $\overline{\Omega}_2$ with the condition*

$$u(x) = v(x), \quad x \in S_0, \tag{4.4}$$

*then the function*

$$w(x) = \begin{cases} u(x), & \\ \quad x \in \Omega_1, & \\ v(x), & \\ \quad x \in \Omega_2, & \\ u(x) = v(x), & \\ \quad x \in S_0, & \end{cases} \tag{4.5}$$

## 4. More Problems for Pluriregular and Plurigeneralized Regular Functions

*is regular or generalized regular in* $\Omega_1 \cup \Omega_2 \cup S_0$. *Moreover, the extension is unique.*

*Proof.* It is sufficient to show that the function (4.5) is regular or generalized regular in the neighborhood of any point of $S_0$. Let $x_0 \in S_0$ and $\Gamma$ be a sphere with the center $x_0$ and with radius $\varepsilon$ so that $S_0 \supset \Gamma \cap S_1 \cap S_2$. Let $\Omega$ be the spherical domain bounded by $\Gamma$ and let $D_1$ and $D_2$ be those parts of this domain that lie in $\Omega_1$ and $\Omega_2$ respectively. Also, let $\Gamma_1 \cup \sigma$, $\Gamma_2 \cup \sigma$ be boundaries of these portions having the common part $\sigma$. Then using (1.76) for generalized regular functions $u(x)$ and $v(x)$ we can write

$$K_{\Gamma_1 \cup \sigma}(u^+, h, x_0) = \frac{1}{2} u^+(x_0),$$

$$K_{\Gamma_2 \cup \sigma}(v^+, h, x_0) = \frac{1}{2} v^+(x_0), \quad x_0 \in S_0.$$

As in (1.67) the normal points outward. In the last two equalities the integrals on $\sigma$ differ only by sign. Therefore, by adding them and taking into consideration (4.4), (4.5) one can get

$$w(x_0) = K_\Gamma(w^+, h, x_0), \quad x_0 \in S_0.$$

Thus the theorem becomes obvious since the right-hand side of this equality represents a generalized regular function inside $\Gamma$. The method of proof used here is well known for holomorphic functions of one complex variable [Mu1].

From this theorem the next result easily follows.

**Generalized Riemann–Schwartz principle of reflection.** Let $\Omega^+$ be the half space $x_n > 0$ with boundary $S: x_n = 0$ and $\Omega^-$ the half space $x_n < 0$. Consider the symmetric point of $x \in \Omega^+$ with respect to $S$

$$x^* = x_0 e_0 + x_1 e_1 + \cdots + x_{n-1} e_{n-1} - x_n e_n. \tag{4.6}$$

We can see that if $u(x)$ is a generalized regular function in $\Omega^+$, i.e., $h$-regular, then

$$u^*(x) = \sum_A u_A(x^*) e_A - \sum_A u_{An}(x^*) e_A e_n, \quad x \in \Omega^-, \tag{4.7}$$

where $A: (\alpha_1, \ldots, \alpha_k)$, $0 \le \alpha_1 < \alpha_2 < \cdots < \alpha_k \le n-1$, is generalized regular in $\Omega^-$ with

$$h^* = \sum_{j=0}^{n-1} h_j e_j - h_n e_n, \quad h = \sum_0^n h_j e_j. \tag{4.8}$$

In fact this follows immediately from the representation

$$\bar{\partial} u + \tilde{u} h = M_1 + M_2,$$

where

$$M_1 = \sum_{A \neq n, k=0}^{n-1} \left[ \frac{\partial u_A(x)}{\partial x_k} e_k e_A + u_A(x) h_k \tilde{e}_A e_k \right] +$$

$$+ \sum_{A \neq n} \left[ \frac{\partial u_{An}(x)}{\partial x_n} e_n e_A e_n + u_{An} h_n \tilde{e}_A \right],$$

$$M_2 = \sum_{A \neq n, k=0}^{n-1} \left[ \frac{\partial u_{An}(x)}{\partial x_k} e_k e_A e_n + u_{An}(x) h_k \tilde{e}_A e_k e_n \right] +$$

$$+ \sum_{A \neq n} \left[ \frac{\partial u_A(x)}{\partial x_n} e_n e_A + u_A h_n \tilde{e}_A e_n \right].$$

Because $[u^*(x)]^* = u(x)$, $u(x)$ will be assumed to take a definite limiting value $u^+(x)$ as $x \to x' \in S$, $x \in \Omega^+$, $[u^*(x)]^-$ also exists because when $x \to x'$, $x \in \Omega^-$, the point $x^* \to x'$, $x^* \in \Omega^+$.

**Theorem.** *Let $u(x)$ be an h-regular function in the domain $D^+$ that lies in $\Omega^+$ and whose boundary $S_0$ contains part of $x_n = 0$. On $S_0$ the following conditions are assumed*

$$u_{An}(x) = 0, \quad A = (\alpha_1, \alpha_2, \ldots, \alpha_k), \quad 0 \leq \alpha_1 < \alpha_2 < \cdots < \alpha_k \leq n-1. \quad (4.9)$$

*Then assuming $h_n = 0$, the function*

$$v(x) = \begin{cases} u(x), & x \in D^+, \\ u^*(x), & x \in D^-, \end{cases} \quad (4.10)$$

*is an h-regular function in the domain $D^+ \cup D^- \cup S_0$, where $D^-$ is a domain symmetric to $D^+$ with respect to $x_n = 0$.*

Since from conditions (4.9) it follows that $u^+(x) = [u^*(x)]^-$, $x \in S_0$, by the extension theorem this theorem is true too.

Suppose that in place of (4.9) we have the conditions

$$u_A(x) = 0, \quad x \in S_0, \quad 0 \leq \alpha_1 < \cdots < \alpha_k \leq n-1. \quad (4.11)$$

## 4. More Problems for Pluriregular and Plurigeneralized Regular Functions

Then the function

$$v(x) = \begin{cases} u(x), & x \in D^+, \\ -u^*(x), & x \in D^-, \end{cases} \quad (4.12)$$

is $h$-regular ($h_n = 0$) in the domain $D^+ \cup D^- \cup S_0$.

The extension theorem for pluriregular or plurigeneralized regular functions is formulated in the following.

**Theorem.** *Let $u(x)$ and $v(x)$ be solutions of equation (4.1) or (4.2) in $\Omega_1$ and $\Omega_2$, respectively, and Hölder-continuous in the closed domains $\Omega_1$ and $\Omega_2$ with the conditions*

$$\overline{\partial}^k u(x) = \overline{\partial}^k v(x), \quad x \in S_0, \quad k = 0, 1, \ldots, m-1. \quad (4.13)$$

*Then the function*

$$w(x) = \begin{cases} u(x), & x \in \Omega_1, \\ v(x), & x \in \Omega_2, \\ u(x) = v(x), & x \in S_0, \end{cases}$$

*is a pluriregular or plurigeneralized regular function in $\Omega_1 \cup \Omega_2 \cup S_0$. Moreover, the extension is unique.*

This theorem is obviously true, like the theorem proved above. Using the Cauchy integral formula (1.67) one can also easily prove that plurigeneralized regular functions in $\Omega \in R^{n+1}$ have derivatives of any order at any internal point.

Now for the solutions of (4.1) and (4.2), the Hilbert and Compound BVP will be considered. Let $S$ be a set of finitely many piecewise-smooth Lyapunov surfaces.

**Hilbert problem.** Find the solution of equation (4.1) or (4.2) with jump surface $S$, that vanishes at infinity by the conditions

$$(\overline{\partial}^k u)^+ - (\overline{\partial}^k u)^- = g_k(x), \quad x \in S, \quad k = 0, 1, \ldots, m-1, \quad (4.14)$$

where $g_k(x)$ are given Hölder-continuous functions.

**Compound BVP.** Let $\Omega^+$ be the half space $x_n > 0$ and $S$ be the piecewise Lyapunov surfaces in $\Omega^+$. Find in $\Omega$ a piecewise pluriregular function with jump surface $S$ that vanishes at infinity by the conditions

$$[\partial^p u(x)]^+ - [\partial^p u(x)]^- = g_p(x), \quad x \in S, \quad p = 0, 1, \ldots, m-1,$$
$$\text{Re}[\partial^p u(x) e_A] = f_A(x), \quad x_n = 0, \quad p = 0, 1, \ldots, m-1,$$

where $A$ takes $2^{n-1}$ different values from $(\alpha_1, \ldots, \alpha_k)$, $0 \leq \alpha_1 < \cdots < \alpha_k \leq n$. These problems can be solved as in the case $m = 1$ in [Ob1].

# Part II

# Initial Value Problems for Regular and Pluriregular, Hyperbolic and Parabolic Equations

*Mathematics is an organism for whose vital strength the indissoluble union of its parts is a necessary condition.*

David Hilbert

# III

# Hyperbolic and Plurihyperbolic Equations in Clifford Analysis

## 0  Introduction

Systems of hyperbolic equations such as Maxwell's equations of electromagnetic fields, the Dirac equations of relativistic quantum mechanics and their generalizations will be considered. They are related to the wave and Klein–Gordon equations.

First, consider the wave equation

$$\Delta u = \frac{\partial^2 u}{\partial t^2}, \quad x_n \equiv t, \quad u = u(x,t), \quad t > 0, \quad x \in R^n, \quad n \geq 1. \tag{1}$$

**Cauchy's problem.** Find the solution $u(x,t)$, a function of the class $C^2(t > 0) \cap C^1(t \geq 0)$, with initial conditions

$$u(x,0) = \phi(x), \quad \left.\frac{\partial u}{\partial t}\right|_{t=0} = \psi(x), \quad x = (x_0, x_1, \ldots, x_{n-1}), \tag{2}$$

where $\phi, \psi \in L(R^n)$ (i.e., $\phi, \psi$ are absolutely integrable).

**Solution.** Without loss of generality one can consider $u(x,0) = 0$. Indeed, let $u_1(x,t), u_2(x,t)$ be the solution of (1) with the conditions

$$\begin{aligned} u_1(x,0) &= 0, \quad \left.\frac{\partial u_1}{\partial t}\right|_{t=0} = \psi(x), \quad x \in R^n, \\ u_2(x,0) &= 0, \quad \left.\frac{\partial u_2}{\partial t}\right|_{t=0} = \phi(x). \end{aligned} \tag{3}$$

Then it is easy to see that the function

$$u(x,t) = u_1(x,t) + \frac{\partial u_2}{\partial t}, \qquad (4)$$

is the solution that meets conditions (2) because, by the condition $u_2(x,0) = 0$ and (1), one has $\left.\frac{\partial^2 u_2}{\partial t^2}\right|_{t=0} = 0$. Thus we can consider equation (1) with the conditions (3).

The FIT with respect to the variables $x_0, x_1, \ldots, x_{n-1}$ of equation (1) and conditions (3) give

$$\frac{d^2 \widehat{u}(y,t)}{dt^2} + |y|^2 \widehat{u}(y,t) = 0, \quad y = (y_0, y_1, \ldots, y_{n-1}),$$
$$\widehat{u}(y,0) = 0, \quad \left.\frac{d\widehat{u}}{dt}\right|_{t=0} = \widehat{\psi}(y), \qquad (5)$$

where $\widehat{u}(y,t)$ is defined by (16) of Chapter II of Part I. The solution of (5) can be represented as

$$\widehat{u}(y,t) = \widehat{\psi}(y) \frac{\sin|y|t}{|y|}.$$

To use the inversion theorem, $u(x,t)$ is represented in the form

$$u(x,t) = \lim_{\varepsilon \to 0} \frac{1}{(\sqrt{2\pi})^n} \int_{R^n} \widehat{\psi}(y) \frac{\sin|y|t}{|y|} e^{-\varepsilon|y|} e^{i(x\cdot y)} dy. \qquad (6)$$

To represent this integral as a convolution, we need to calculate the integral

$$J = \frac{1}{(\sqrt{2\pi})^n} \int_{R^n} \widehat{\psi}(y) \frac{\sin|y|t}{|y|} e^{-\varepsilon|y|+i(x\cdot y)} dy, \quad \varepsilon > 0.$$

Let $n = 2m$. Then by (35), (36) of Chapter II of Part I we have

$$J \equiv J_1 = (-2)^{m-1} \frac{d^{m-1}}{d(\rho^2)^{m-1}} \int_0^\infty e^{-\varepsilon r} \sin rt \, J_0(r\rho) \, dr, \quad \rho = |x|, \quad r = |y|,$$

and for $n = 2m+1$

$$J \equiv J_2 = -(-2)^m \frac{1}{\sqrt{2\pi}} \frac{d^{m-1}}{d(\rho^2)^{m-1}} \frac{1}{\rho} \int_0^\infty e^{-\varepsilon r} \sin rt \sin r\rho \, dr. \qquad (7)$$

## 0. Introduction

It is known that

$$\int_0^\infty e^{-\varepsilon r} \sin rt \, J_0(r\rho) \, dr = -I_m\left[\frac{1}{\sqrt{\rho^2 + a^2}}\right], \quad a = \varepsilon + it.$$

Thus

$$J_1 = -(-2)^{m-1} I_m\left[\frac{d^{m-1}}{d(a^2)^{m-1}} \frac{1}{\sqrt{a^2 + \rho^2}}\right],$$

where $\lim_{\varepsilon \to 0} J_1 = 0$ for $\rho > t$, and

$$\lim_{\varepsilon \to 0} J_1 = 2^{m-1} \frac{d^{m-1}}{d(t^2)^{m-1}} \frac{1}{\sqrt{t^2 - \rho^2}}, \quad \rho < t,$$

or

$$\lim_{\varepsilon \to 0} J_1 = (-2)^{m-1} \frac{d^{m-1}}{d(\rho^2)^{m-1}} \frac{1}{\sqrt{t^2 - \rho^2}}, \quad \rho < t. \quad (8)$$

Thus, for $n = 2m$, (6) can be represented as

$$u(x, t) = \frac{1}{2\pi^m} \frac{d^{m-1}}{d(t^2)^{m-1}} \left[t^{n-1} \int_{|y| \le 1} \frac{\psi(x - ty)}{\sqrt{1 - |y|^2}} \, dy\right]. \quad (9)$$

Now for $J_2$, we can write

$$J_2 = \frac{(-2)^{m-1}}{\sqrt{2\pi}} \frac{d^{m-1}}{d(\rho^2)^{m-1}} \frac{1}{\rho} \int_0^\infty e^{-\varepsilon r}\left[\cos r(\rho - t) - \cos r(\rho + t)\right] dr.$$

This integral is calculated very easily and we can write

$$J_2 = \frac{(-2)^{m-1}}{\sqrt{2\pi}} \frac{d^{m-1}}{d(\rho^2)^{m-1}} \frac{1}{\rho}\left[\frac{\varepsilon}{\varepsilon^2 + (\rho - t)^2} - \frac{\varepsilon}{\varepsilon^2 + (\rho + t)^2}\right], \quad (10)$$

so that for $n = 2m + 1$, (6) can be represented as

$$u(x, t)$$
$$= \frac{(-2)^{m-1}}{(\sqrt{2\pi})^{n+1}} \lim_{\varepsilon \to 0} \int_{R^n} \psi(x - y) \frac{d^{m-1}}{d(r^2)^{m-1}} \frac{1}{r}\left[\frac{\varepsilon}{\varepsilon^2 + (r - t)^2} - \frac{\varepsilon}{\varepsilon^2 + (r + t)^2}\right] dy.$$

It is clear that the limit of the second addend is zero but that the limit of the first addend is not zero. Using integration by parts, the limit of the first addend can be rewritten in the form

$$u(x,t) = \frac{1}{4\pi^{m+1}} \lim_{\varepsilon \to 0} \int_{|y|=1} ds_y \int_0^\infty \frac{d^{m-1}}{d(r^2)^{m-1}} [r^{n-2}\psi(x-y)] \frac{\varepsilon dr}{\varepsilon^2 + (r-t)^2}.$$

Changing the integration variable $r = t + \varepsilon\tau$ one can get

$$u(x,t) = \frac{1}{4\pi^m} \frac{d^{m-1}}{d(t^2)^{m-1}} \left[ t^{n-2} \int_{|y|=1} \psi(x-ty) ds_y \right]. \tag{11}$$

For $u(x,t)$ to be a regular solution one can suppose that $\psi(x)$ is a function with continuous partial derivatives to order $(n+1)/2$ when $n$ is odd and to order $(n+2)/2$ when $n$ is even.

**Theorem.** *If $\psi(x)$ is odd or even with respect to $x_{n-1}$, then for (9) as well as for (11), the following equalities are true, respectively*

$$\lim_{x_{n-1} \to 0} u(x,t) = 0 \quad \text{or} \quad \lim_{x_{n-1} \to 0} \frac{\partial u(x,t)}{\partial x_{n-1}} = 0. \tag{12}$$

The proof follows easily if we consider spherical coordinates (equation (26), Part I, Chapter II). In fact, since $dy = |y|^{n-1} \sin^{n-1}\phi_1 dr\, d\phi_1\, ds_1$, $ds_y = \sin^{n-2}\phi_1 d\phi_1\, d\omega_{n-1}$, $y_{n-1} = |y|\cos\phi_1$, we have

$$\int_0^\pi \psi(x-ty) \sin^{n-1}\phi_1\, d\phi_1 =$$

$$= \int_0^{\frac{\pi}{2}} [\psi(x_0 - ty_0, \ldots, x_{n-1} - ty_{n-1}) +$$

$$+ \psi(x_0 - ty_0, \ldots, x_{n-2} - ty_{n-2}, x_{n-1} + ty_{n-1})] \sin^{n-2}\phi_1\, d\phi_1.$$

Thus if $\psi$ is odd with respect to $x_{n-1}$, then the first equality from (12) is valid, but if $\psi$ is even with respect to $x_{n-1}$, then its derivative with respect to this variable is odd and the second equality from (12) is valid.

Using this theorem we can solve a mixed problem. Find the solution of equation (1) in the half space $x_{n-1} > 0$ for $t > 0$ with the conditions

$$u(x_0, \ldots, x_{n-2}, 0, t) = 0 \quad \text{or} \quad \frac{\partial u}{\partial x_{n-1}} = 0 \quad \text{for} \quad x_{n-1} = 0, \tag{13}$$

and

$$u(x, 0) = 0, \quad \frac{\partial u}{\partial t} = \psi(x) \quad \text{for} \quad t = 0, \quad x_{n-1} > 0. \tag{14}$$

If $\psi(x)$ for $x_{n-1} < 0$ is continued in the odd sense or even sense, this problem can be reduced to Cauchy's problem (3); thus (13), (14) can be solved explicitly too.

Now consider Cauchy's problem for a nonhomogeneous equation. Find the solution of the equation

$$\frac{\partial^2 u}{\partial t^2} = \Delta u + f(x, t), \quad x \in R^n, \quad t > 0, \tag{15}$$

with the conditions

$$u(x, 0) = \frac{\partial u}{\partial t} = 0 \quad \text{for} \quad t = 0. \tag{16}$$

It is easy to obtain the FIT of $u(x, t)$ with respect to $x_0, x_1, \ldots, x_{n-1}$:

$$\widehat{u}(x, t) = \frac{1}{|y|} \int_0^t \widehat{f}(y, \tau) \sin |y|(t - \tau) \, d\tau. \tag{17}$$

By the inversion theorem, this can be represented as in (6), and in light of (9), (11), it can be represented in an analogous way.

Now consider the case

$$f(x, t) = \delta(x) f(t), \tag{18}$$

where $\delta(x)$ is the Dirac delta function, the FIT of which with respect to $x_0, \ldots, x_{n-1}$ equals one. In this case the problem (15), (16) is called the radiation problem [CH]. Then by (17) we have

$$u(x, t) = \lim_{\varepsilon \to 0} \frac{1}{(\sqrt{2\pi})^n} \int_0^t f(t - \tau) d\tau \int_{R^n} e^{-\varepsilon |y| + i(x \cdot y)} \frac{\sin |y| \tau}{|y|} dy.$$

By (8) and (10), this can be written in the form

$$u(x, t) = \frac{(-1)^{m-1}}{2\pi^m} \frac{d^{m-1}}{d(r^2)^{m-1}} \int_r^t \frac{f(t - \tau) d\tau}{\sqrt{\tau^2 - r^2}}, \quad r < t, \quad \text{for} \quad n = 2m, \tag{19}$$

and

$$u(x,t) = \frac{(-1)^{m-1}}{4\pi^{m+1}} \frac{d^{m-1}}{d(r^2)^{m-1}} \frac{1}{r} \lim_{\varepsilon \to 0} \int_0^t f(t-\tau) \frac{\varepsilon d\tau}{\varepsilon^2 + (r-\tau)^2},$$
$$r < t, \quad \text{for } n = 2m+1.$$

In both cases $u(x,t) = 0$ for $r \geq t$, $r = |x|$.

Changing the integrand variable in the last integral to $\tau = r + \varepsilon \tau_1$ gives

$$u(x,t) = \frac{(-1)^{m-1}}{4\pi^m} \frac{d^{m-1}}{d(r^2)^{m-1}} \frac{f(t-r)}{r}, \quad r < t. \tag{20}$$

These formulas were obtained in a different way in [CH].

**Cauchy's problem for the Klein–Gordon equation.** Find the regular solution of the equation

$$\Delta u - h^2 u = \frac{\partial^2 u}{\partial t^2}, \quad x \in R^n, \quad t > 0, \tag{21}$$

with the conditions

$$u(x,0) = 0, \quad \frac{\partial u}{\partial t} = \psi(x) \quad \text{for } t = 0. \tag{22}$$

Equation (21) with conditions (2) can be solved as in (4) if we know its solution with conditions (22).

The FIT of (21) and (22) with respect to the variables $x_0, x_1, \ldots, x_{n-1}$ gives us

$$\widehat{u}(y,t) = \frac{\sin \lambda t}{\lambda} \widehat{\psi}(y), \quad y \in R^n,$$

where $\lambda = \sqrt{|y|^2 + h^2}$. Then, using the inversion theorem, $u(x,t)$ can be written as

$$u(x,t) = \frac{1}{(\sqrt{2\pi})^n} \lim_{\varepsilon \to 0} \int_{R^n} \widehat{\psi}(y) \frac{\sin \lambda t}{\lambda} e^{-\varepsilon \lambda + i(x \cdot y)} dy. \tag{23}$$

To represent this integral as a convolution we need to find the integral

$$J = \frac{1}{(\sqrt{2\pi})^n} \lim_{\varepsilon \to 0} \int_{R^n} e^{-\varepsilon \lambda + i(x \cdot y)} \frac{\sin \lambda t}{\lambda} dy. \tag{24}$$

Let $n = 2m$. Then by (35) of Chapter II of Part I, we have

$$J = (-2)^{m-1} \frac{d^{m-1}}{d(r^2)^{m-1}} \lim_{\varepsilon \to 0} \int_0^\infty e^{-\varepsilon \lambda} \frac{\sin \lambda t}{\lambda} \rho J_0(r\rho) \, d\rho, \quad \rho = |y|, \quad r = |x|,$$

which can be written as

$$J = -(-2)^{m-1} \frac{d^{m-1}}{d(r^2)^{m-1}} \lim_{\varepsilon \to 0} \operatorname{Im} \int_0^\infty e^{-(\varepsilon + it)\lambda} \frac{\rho J_0(r\rho)}{\lambda} \, d\rho.$$

By the Sommerfeld formula (44) of Chapter II of Part I we have

$$J = -(-2)^{m-1} \frac{d^{m-1}}{d(r^2)^{m-1}} \lim_{\varepsilon \to 0} \operatorname{Im} \frac{\exp[-h\sqrt{r^2 + (\varepsilon + it)^2}]}{\sqrt{r^2 + (\varepsilon + it)^2}}.$$

Thus

$$J = \begin{cases} 0, & \text{for } t < r, \\ (-2)^{m-1} \dfrac{d^{m-1}}{d(r^2)^{m-1}} \dfrac{\cos h\sqrt{t^2 - r^2}}{\sqrt{t^2 - r^2}}, & \text{for } t > r, \end{cases}$$

and by (8), (23), $u(x, t)$ has the form

$$u(x, t) = \frac{1}{2\pi^m} \frac{d^{m-1}}{d(t^2)^{m-1}} \int_{|y| \le t} \psi(x - y) \frac{\cos h\sqrt{t^2 - \rho^2}}{\sqrt{t^2 - \rho^2}} \, dy \quad \text{for } n = 2m. \tag{25}$$

Now let $n = 2m + 1$. Then by (36) of Chapter II of Part I and (24), we have

$$J = (-2)^m \sqrt{\frac{2}{\pi}} \frac{d^m}{d(r^2)^m} \lim_{\varepsilon \to 0} \int_0^\infty e^{-\varepsilon \lambda} \frac{\sin \lambda t}{\lambda} \cos r\rho \, d\rho,$$

which can be rewritten as

$$J = (-2)^m \frac{1}{\sqrt{2\pi}} \frac{d^m}{d(r^2)^m} \lim_{\varepsilon \to 0} \operatorname{Im} \int_{-\infty}^\infty e^{-(\varepsilon - it)\lambda + ir\rho} \frac{d\rho}{\lambda}.$$

Using (46) of Chapter II of Part I we have

$$J = (-2)^m \sqrt{\frac{\pi}{2}} \frac{d^m}{d(r^2)^m} \lim_{\varepsilon \to 0} \operatorname{Im} \left[ i H_0^{(1)}\left(ih\sqrt{r^2 + (\varepsilon - it)^2}\right) \right].$$

Taking into consideration the properties of the Hankel function of zeroth order, we have

$$\mathrm{Im}\left[i H_0^{(1)}(ih\sqrt{r^2 - t^2})\right] = \begin{cases} 0, & r > t, \\ J_0(h\sqrt{t^2 - r^2}), & r \le t, \end{cases}$$

where $J_0$ is the Bessel function of zeroth order. Thus, by (8)

$$J = (-2)^m \sqrt{\frac{\pi}{2}} \frac{d^m}{d(t^2)^m} J_0(h\sqrt{t^2 - r^2}) \quad \text{for } r \le t,$$

and $u(x, t)$ can be represented as the convolution

$$u(x, t) = \frac{1}{2\pi^m} \frac{d^m}{d(t^2)^m} \int_{|y| \le t} \psi(x - y) J_0\left(h(\sqrt{t^2 - |y|^2})\right) dy, \quad n = 2m + 1. \tag{26}$$

To explicitly solve the radiation problem for the equation

$$\frac{\partial^2 u}{\partial t^2} = \Delta u - h^2 u + f(x, t),$$

with conditions (26), (18) we only need to repeat the method used for the solution of the problem for equation (15).

For the representations (25), (26) the theorem with (12) is true as proved for the representations (9), (11). So the mixed problem for equation (21) with conditions (13), (14) can be solved too.

From the representation (11) the so-called Huygens' principle follows. If $n > 1$ is odd, then the solution of the Cauchy problem for the wave equation depends only on the initial functions given only on $|y| = t$, i.e., on the base of the characteristic cone, but it does not depend on the initial functions given inside this base. If $n$ is even, from (9) it follows that Huygens' principle has no place. Thus for $n$ odd, if the initial functions are nonzero only locally, then their behavior at every point in space is local in time. Because $n$ is the number of space variables and because there is one time variable, in our world Huygens' principle is true and the spreading of waves has many useful applications.

# 1 IVP for Hyperbolic Systems (Maxwell and Dirac Equations)

Consider the Clifford space $R_{(n,n-1)}$ and the operators

# 1. IVP for Hyperbolic Systems (Maxwell and Dirac Equations)

$$\overline{\partial} = \sum_0^n \frac{\partial}{\partial x_k} e_k, \quad \partial = \frac{\partial}{\partial x_0} e_0 - \sum_1^n \frac{\partial}{\partial x_k} e_k, \quad x = \sum_0^n x_k e_k, \tag{1.1}$$

where $e_k$ ($k = 0, \ldots, n$) have properties,

$$\begin{gathered} e_0^2 = e_0, \quad e_k^2 = -e_0, \quad k = 1, \ldots, n-1, \quad e_n^2 = e_0, \\ e_i e_k + e_k e_i = 0, \quad k \ne i, \quad k, i = 1, 2, \ldots, n. \end{gathered} \tag{1.2}$$

Let $u(x)$ be an element of $R_{(n,n-1)}$ and a solution of the equation

$$\overline{\partial} u = 0. \tag{1.3}$$

**Cauchy's problem.** Find a regular solution $u(x, t)$ of (1.3) with values in $R_{(n,n-1)}$ for $x = (x_0, \ldots, x_{n-1}) \in R^n$ and $x_n \equiv t \ge 0$ subject to the condition

$$u(x, 0) = \varphi(x), \tag{1.4}$$

where the given function $\varphi(x)$ with values in $R_{(n,n-1)}$ has continuous partial derivatives of the necessary order.

**Solution.** From (1.4) the quantities

$$\frac{\partial u}{\partial x_k} = \frac{\partial \varphi}{\partial x_k} \quad \text{for} \quad t = 0, \quad k = 0, 1, \ldots, n-1,$$

are prescribed. From equation (1.3) we can derive

$$\frac{\partial u}{\partial t} = -e_n \sum_{k=0}^{n-1} e_k \frac{\partial u}{\partial x_k} \quad \text{for} \quad t = 0. \tag{1.5}$$

Because $u$ is also a solution of the wave equation, by Cauchy's condition (1.4), (1.5) is represented explicitly in quadratures by (9), (11). Now we prove that $u$ solves (1.3).

It is obvious that if $u$ solves the wave equation, then so does $\overline{\partial} u$. From (1.3) and (1.5) we see that

$$\begin{gathered} \overline{\partial} u = 0 \quad \text{for} \quad t = 0, \\ \frac{\partial^2 u}{\partial t \partial x_0} = -e_n \sum_{k=0}^{n-1} e_k \frac{\partial^2 u}{\partial x_k \partial x_0} \quad \text{for} \quad t = 0, \\ \sum_{j=1}^{n-1} e_j \frac{\partial^2 u}{\partial t \partial x_j} = e_n \sum_{j=1}^{n-1} \sum_{k=0}^{n-1} e_j e_k \frac{\partial^2 u}{\partial x_k \partial x_j} \quad \text{for} \quad t = 0. \end{gathered} \tag{1.6}$$

This last equality can be written as

$$\sum_{j=1}^{n-1} e_j \frac{\partial^2 u}{\partial t \partial x_j} = e_n \sum_{j=1}^{n-1} e_j \frac{\partial^2 u}{\partial x_0 \partial x_j} - e_n \sum_{k=1}^{n-1} \frac{\partial^2 u}{\partial x_k^2} \quad \text{for} \quad t = 0.$$

Then we have

$$\sum_{j=0}^{n-1} e_j \frac{\partial^2 u}{\partial t \partial x_j} = -e_n \Delta u = -e_n \frac{\partial^2 u}{\partial t^2} \quad \text{for} \quad t = 0,$$

and consequently,

$$\frac{\partial(\bar{\partial}u)}{\partial t} = 0 \quad \text{for} \quad t = 0. \tag{1.7}$$

Because Cauchy's problem for the wave equation has a unique solution, from (1.6) and (1.7) it follows that

$$\bar{\partial}u = 0 \quad \text{for} \quad x \in R^n, \quad t > 0.$$

Now consider the nonhomogeneous equation

$$\bar{\partial}u = f(x, t), \quad x \in R^n, \quad t \geq 0, \tag{1.8}$$

with the condition

$$u(x, 0) = 0, \tag{1.9}$$

where $f$ is a given function with values in $R_{(n,n-1)}$. This problem can be easily reduced to Cauchy's problem for the homogeneous equation (1.3). Indeed, let $v(x, t, \tau)$ for $t > \tau$ be the solution of (1.3) with the condition

$$v(x, \tau, \tau) = e_n f(x, \tau), \quad e_n^2 = e_0.$$

Then the solution of (1.8), (1.9) is given as

$$u(x, t) = \int_0^t v(x, t, \tau) d\tau,$$

## 1. IVP for Hyperbolic Systems (Maxwell and Dirac Equations)

because

$$\bar{\partial}u = \int_0^t \bar{\partial}v(x, t, \tau)\, d\tau + e_n v(x, t, t).$$

Consider the hyperbolic $h$-regular equation

$$\bar{\partial}u + \tilde{u}h = 0, \tag{1.10}$$

where $h$ is a vectorial constant. Then $u(x)$ is also the solution of the Klein–Gordon equation

$$\Delta u - |h|^2 u = \frac{\partial^2 u}{\partial x_n^2}, \quad n \geq 1, \quad x_n \equiv t \geq 0. \tag{1.11}$$

**Cauchy's IVP.** Find a regular solution $u(x, t)$ of (1.10) with values in $R_{(n,n-1)}$ for $x \in R^n$ subject to the condition

$$u(x, 0) = \varphi(x), \quad x(x_0, \ldots, x_{n-1}). \tag{1.12}$$

**Solution.** By (1.12) the following quantities are given

$$\frac{\partial u}{\partial x_k} = \frac{\partial \varphi}{\partial x_k} \quad \text{for} \quad t = 0, \quad k = 0, 1, \ldots, n-1.$$

Then from (1.10) it follows that

$$e_n \frac{\partial u}{\partial t} = -\sum_{k=0}^{n-1} e_k \frac{\partial \varphi}{\partial x_k} - \tilde{\varphi}h \quad \text{for} \quad t = 0. \tag{1.13}$$

Because $u$ is also the solution of (1.11), by (1.12), (1.13) it is represented explicitly in the form (25), (26).

If we consider the nonhomogeneous equation

$$\bar{\partial}u + \tilde{u}h = f(x, t), \tag{1.14}$$

with the condition

$$u(x, 0) = 0,$$

136    III. Hyperbolic and Plurihyperbolic Equations in Clifford Analysis

the solution can be defined again by the representation

$$u(x, t) = \int_0^t v(x, t, \tau) \, d\tau,$$

where $v(x, t, \tau)$, $t > \tau$, is the solution of (1.10) with the condition

$$v(x, \tau, \tau) = e_n f(x, \tau).$$

Now consider the Clifford algebra $R_{(3,2)}$, which is especially interesting because equation (1.3) represents the Dirac equations of relativistic quantum mechanics and, in a particular case, the Maxwell equations of electromagnetic fields.

Let $u(x, t) \in R_{(3,2)}$, $x_3 \equiv t$, and

$$u(x) = u_0 e_0 - u_1 e_1 - u_2 e_2 - \Phi e_3 - \Psi e_1 e_2 - u_3 e_3 e_1 - u_4 e_2 e_3 - u_5 e_1 e_2 e_3. \tag{1.15}$$

Then (1.3) gives

$$\operatorname{div} E - \frac{\partial \Phi}{\partial t} = 0, \quad \operatorname{div} H + \frac{\partial \Psi}{\partial t} = 0,$$
$$\operatorname{grad} \Psi + \operatorname{rot} E + \frac{\partial H}{\partial t} = 0, \quad \operatorname{grad} \Phi + \operatorname{rot} H - \frac{\partial E}{\partial t} = 0, \tag{1.16}$$

where $E = (u_0, u_1, u_2)$, $H = (u_5, u_4, u_3)$ and the operators div, grad, rot are taken with respect to $x_0, x_1, x_2$. If $\Phi = \Psi \equiv 0$, we have the Maxwell equations

$$\operatorname{div} E = 0, \quad \operatorname{div} H = 0,$$
$$\operatorname{rot} E + \frac{\partial H}{\partial t} = 0, \quad \operatorname{rot} H - \frac{\partial E}{\partial t} = 0. \tag{1.17}$$

In the case of the $h$-regular equation (1.10) for $u(x, t) \in R_{(3,2)}$ in the form (1.15), we have the generalization of the Dirac equations (1.16)

$$\operatorname{div} E - \frac{\partial \Phi}{\partial t} + (E \cdot A) - \Phi h_3 = 0,$$
$$\operatorname{grad} \Psi + \operatorname{rot} E + \frac{\partial H}{\partial t} + [E \times A] + \Psi A + H h_3 = 0,$$
$$\operatorname{div} H + \frac{\partial \Psi}{\partial t} - (H \cdot A) - \Psi h_3 = 0,$$
$$\operatorname{grad} \Phi + \operatorname{rot} H - \frac{\partial E}{\partial t} - [H \times A] + \Phi A + E h_3 = 0, \tag{1.18}$$

where $A$ is the three-component vector $(h_0, h_1, h_2)$, $h = h_0 e_0 - \sum_1^3 h_k e_k$. If $\Phi = \Psi \equiv 0$, we will have from (1.18) the generalized Maxwell equations. Cauchy's problem for all these equations are solved above.

Let us consider equation (1.3) for vectorial $u(x) \in R_{(n,n-1)}$

$$u(x) = u_0 e_0 - \sum_1^n u_k e_k. \tag{1.19}$$

This equation is equivalent to the system

$$\sum_0^{n-1} \frac{\partial u_k}{\partial x_k} - \frac{\partial u_n}{\partial x_n} = 0,$$
$$\frac{\partial u_i}{\partial x_k} - \frac{\partial u_k}{\partial x_i} = 0, \quad i, k = 0, 1, \ldots, n. \tag{1.20}$$

For $n > 1$, this is an overdetermined hyperbolic system, and for $n = 1$, it is the simplest classical hyperbolic system. System (1.20) can be considered the hyperbolic analogue of the Riesz system in the elliptic case.

Moreover, equations (1.10) with (1.19) give the hyperbolic analogue of the generalized Riesz system in the elliptic case

$$\sum_0^{n-1} \left[ \frac{\partial u_k}{\partial x_k} + h_k u_k \right] - \frac{\partial u_n}{\partial x_n} - h_n u_n = 0,$$
$$\frac{\partial u_j}{\partial x_k} - \frac{\partial x_k}{\partial x_j} - h_k u_j + h_j u_k = 0, \quad j, k = 0, 1, \ldots, n, \tag{1.21}$$

where $h = \sum_0 h_k e_k$. For these equations Cauchy's problem can be formulated and solved as above for equation (1.3) with condition (1.4).

## 2 B&IVP for Pluriregular and Plurigeneralized Regular Hyperbolic Systems, Polywave and Poly-Klein–Gordon Equations, Harmonic-Wave and Harmonic Klein–Gordon Equations

Let $u(x, t) \in R_{(n,n-1)}$ and consider the high-order equation

$$\overline{\partial}^m u = 0, \quad m \geq 1. \tag{2.1}$$

It is clear that $u(x, t)$ is also a solution of the polywave equation

$$\left(\Delta - \frac{\partial^2}{\partial t^2}\right)^m u = 0. \tag{2.2}$$

If $u_0, \ldots, u_m$ are solutions of the wave equation, then

$$u = \sum_0^{m-1} t^k u_k$$

is a solution of the polywave equation (2.2). An analogous formula was used to solve BVP for polyharmonic functions. But this representation cannot be used successfully to solve IVP.

For $m = 2$ (2.1) is called a biregular system and (2.2) the biwave equation. Consider a biwave equation of the form

$$\left(\Delta - a^2 \frac{\partial^2}{\partial t^2}\right)\left(\Delta - b^2 \frac{\partial^2}{\partial t^2}\right) u = 0, \tag{2.3}$$

which has applications in the theory of elasticity. The moving equations of isotropic elastic bodies in displacement coordinates are [Mu2]:

$$(\lambda + \mu)\frac{\partial \theta}{\partial x_k} + \mu \Delta u_k = \rho \frac{\partial^2 u_k}{\partial t^2}, \quad k = 0, 1, 2,$$
$$\theta = \frac{\partial u_0}{\partial x_0} + \frac{\partial u_1}{\partial x_1} + \frac{\partial u_2}{\partial x_2}. \tag{2.4}$$

From (2.4) it follows that

$$\Delta \theta = \frac{\rho}{\lambda + 2\mu} \frac{\partial^2 \theta}{\partial t^2},$$
$$(\lambda + \mu)\frac{\partial \Delta \theta}{\partial x_k} + \mu \Delta \Delta u_k - \rho \frac{\partial^2 \Delta u_k}{\partial t^2} = 0, \tag{2.5}$$
$$(\lambda + \mu)\frac{\partial^3 \theta}{\partial x_k \partial t^2} + \mu \frac{\partial^2 \Delta u_k}{\partial t^2} = \rho \frac{\partial^4 u_k}{\partial t^4}.$$

In other words, one has

$$\Delta \Delta u_k - \frac{\rho}{\mu} \frac{\lambda + 3\mu}{\lambda + 2\mu} \frac{\partial^2 \Delta u_k}{\partial t^2} + \frac{\rho^2}{\mu(\lambda + 2\mu)} \frac{\partial^4 u_k}{\partial t^4} = 0,$$

which can be written as

$$\left(\Delta - a^2 \frac{\partial^2}{\partial t^2}\right)\left(\Delta - b^2 \frac{\partial^2}{\partial t^2}\right) u = 0,$$

where $a^2 = \frac{\rho}{\mu}$, $b^2 = \frac{\rho}{\lambda + 2\mu}$.

**Cauchy's problem for (2.3).** Find the regular solution $u(x, t)$ of equation (2.3) for $x(x_0, \ldots, x_{n-1}) \in R^n$, $t \equiv x_n \geq 0$, with the conditions

$$u(x, 0) = f_0(x), \quad \left.\frac{\partial u}{\partial t}\right|_{t=0} = f_1(x), \quad \left.\frac{\partial^2 u}{\partial t^2}\right|_{t=0} = f_2(x), \quad \left.\frac{\partial^3 u}{\partial t^3}\right|_{t=0} = f_3(x). \tag{2.6}$$

**Solution.** By these conditions, we can define for $t = 0$

$$v(x, t) \equiv \Delta u - b^2 \frac{\partial^2 u}{\partial t^2} \quad \text{and} \quad \frac{\partial v}{\partial t} = \frac{\partial \Delta u}{\partial t} - b^2 \frac{\partial^3 u}{\partial t^3}.$$

Thus for the equation

$$\Delta v - a^2 \frac{\partial^2 v}{\partial t^2} = 0,$$

we have Cauchy's problem, the solution of which is given as (9), (11). Then for the equation

$$\Delta u - b^2 \frac{\partial^2 u}{\partial t^2} = v(x, t), \tag{2.7}$$

we have Cauchy's problem, and its solution is once again represented in quadratures. In that way, Cauchy's problem can be solved for the equation

$$\left(\Delta - a_1^2 \frac{\partial^2}{\partial t^2}\right) \cdots \left(\Delta - a_m^2 \frac{\partial^2}{\partial t^2}\right) u = 0, \tag{2.8}$$

with the conditions for $t = 0$

$$u(x, 0) = f_0(x), \quad \frac{\partial u}{\partial t} = f_1(x), \ldots, \frac{\partial^{2m-1} u}{\partial t^{2m-1}} = f_{2m-1}. \tag{2.9}$$

Now consider (2.1) in the case $m = 2$.

140     III. Hyperbolic and Plurihyperbolic Equations in Clifford Analysis

**Cauchy's problem for (2.1).** Let $u(x,t) \in R_{(n,n-1)}$. Find the regular solution of (2.1) with the conditions for $t=0$

$$u(x,0) = f_0(x), \quad \frac{\partial u}{\partial t} = f_1(x). \qquad (2.10)$$

**Solution.** By these conditions one can define $\bar{\partial} u$ for $t=0$. Thus we have Cauchy's problem (1.4) for equation (1.3), which is solved. Then we have the nonhomogeneous equation (1.8) with the condition $u(x,0) = f_0(x)$. Represent the solution in the form

$$u(x,t) = \overset{1}{u}(x,t) + \overset{2}{u}(x,t),$$

where $\overset{1}{u}(x,t)$ is the solution of homogeneous equation (1.3) with nonhomogeneous condition $\overset{1}{u}(x,0) = f_0(x)$, and $\overset{2}{u}(x,t)$ is the solution of (1.8) with homogeneous condition $\overset{2}{u}(x,0) = 0$. Both $\overset{1}{u}, \overset{2}{u}$ can be defined as the solutions of problems (1.3), (1.4) and (1.8), (1.9) are defined.

Now consider the plurigeneralized regular equation of $m$-th order

$$P^m u = 0, \quad Pu = \bar{\partial} u + \tilde{u} h, \quad h = \sum_0^n h_k e_k. \qquad (2.11)$$

In the case $m=2$, we have the bigeneralized regular equation, which can be written as

$$\bar{\partial}(\bar{\partial} u + \tilde{u} h) + (\partial \tilde{u} + uh)h = 0. \qquad (2.12)$$

By force of (1.11), $u(x,t)$ is also a solution of the bi-Klein–Gordon equation

$$\left(\Delta - |h|^2 - \frac{\partial^2}{\partial t^2}\right)^2 u = 0. \qquad (2.13)$$

If $m > 2$, one can obtain the poly-Klein–Gordon equation

$$\left(\Delta - |h|^2 - \frac{\partial^2}{\partial t^2}\right)^m u = 0. \qquad (2.14)$$

Cauchy's problem for these equations will be posed as (2.6) and (2.9), and using (25), (26), the solution will be represented in quadratures. Cauchy's problem for equation (2.12) with conditions (2.10) will also be solved as (2.10) for equation (2.1). For any $m$, one can solve Cauchy's problem in the same way.

## 2. B&IVP for ... and Harmonic Klein Gordon Equations

Now consider the equation

$$\bar{\partial}^m \left( \bar{\partial} + \frac{\partial}{\partial t} e_n \right)^m u(x,t) = 0, \quad m \geq 1, \tag{2.15}$$

where

$$\bar{\partial} = \sum_0^{n-1} \frac{\partial}{\partial x_k} e_k, \quad e_k^2 = -e_0, \quad k = 1, \ldots, n-1, \quad e_n^2 = e_0.$$

This equation is the plurielliptic-plurihyperbolic equation. It is clear that $u(x,t)$ is also the solution of the equation

$$\Delta^m \left( \Delta - \frac{\partial^2}{\partial t^2} \right)^m u(x,t) = 0,$$

which can be called the polyharmonic-polywave equation. It will be interesting to consider $m = 1$, i.e.,

$$\Delta \left( \Delta - \frac{\partial^2}{\partial t^2} \right) u(x,t) = 0, \quad x(x_0, \ldots, x_{n-1}), \tag{2.16}$$

called the harmonic-wave equation. The following problems are correctly posed and will be solved in quadratures.

**Dirichlet–Cauchy problem.** Find the regular solution of (2.16) for $t > 0$, $x_{n-1} > 0$, that vanishes at infinity by the conditions

$$u(x,0) = \varphi_1(x), \quad \frac{\partial u}{\partial t} = \varphi_2(x), \quad t = 0, \tag{2.17}$$

$$u(x,t) = \varphi(x_0, \ldots, x_{n-2}, t), \quad x_{n-1} = 0, \quad t > 0, \tag{2.18}$$

with the corresponding compatibility conditions for these given functions.

**Solution.** Let

$$\Delta u(x,t) = F(x,t), \tag{2.19}$$

$$\Delta F - \frac{\partial^2 F}{\partial t^2} = 0. \tag{2.20}$$

Then by force of (2.17), (2.19), the unknown function $F$ satisfies

$$F(x,0) = \Delta \varphi_1(x) \equiv f_1(x), \quad \frac{\partial F}{\partial t} = \Delta \varphi_2(x) \equiv f_2(x), \quad t = 0;$$

i.e., to define $F$, we have Cauchy's IVP for the wave equation, which was treated in the introduction.

To define $u(x, t)$, we have the Dirichlet problem for the nonhomogeneous equation (2.19) with the condition (2.18). The solution is given in the first part.

**Neumann–Cauchy problem.** Find the solution of (2.16) for $t > 0$, $x_{n-1} > 0$, that vanishes at infinity by conditions (2.17) and

$$\frac{\partial u}{\partial x_{n-1}} = \varphi(x_0, \ldots, x_{n-2}, t), \quad x_{n-1} = 0, \quad t > 0. \tag{2.21}$$

The solution can be reduced to that of the Neumann problem for equation (2.19) and represented in quadratures, as given in the first part.

All problems considered for harmonic functions in the first part can be correspondingly considered here. For instance, when $x(x_0, x_1, x_2)$ belongs to three-dimensional space with a crack along $x_0^2 + x_1^2 \leq a^2$, $x_2 = 0$, the solution can be represented effectively using Hobson's formula. In the same way, the harmonic Klein–Gordon equation

$$\Delta\left(\Delta - k^2 - \frac{\partial^2}{\partial t^2}\right)u(x, t) = 0,$$

can be considered, for which problems (2.17), (2.18) or (2.21) can be solved.

This equation is connected with the elliptic regular and hyperbolic generalized regular equation

$$\overline{\partial}\left(\overline{\partial}u + e_n \frac{\partial u}{\partial t} + \widetilde{u}h\right) = 0.$$

Moreover, the Helmholtz-wave equation or Helmholtz–Klein–Gordon equation

$$(\Delta - k_1^2)\left(\Delta - \frac{\partial^2}{\partial t^2}\right)u(x, t) = 0,$$

$$(\Delta - k_1^2)\left(\Delta - k_2^2 - \frac{\partial^2}{\partial t^2}\right)u(x, t) = 0,$$

can be considered, for which problems (2.17), (2.18) or (2.21) are correctly posed and solvable in quadratures too.

For equation (2.15), boundary-initial value problems (B&IVP) can be considered correspondingly. For instance, in the case $m = 1$

$$\overline{\partial}\left(\overline{\partial} + \frac{\partial}{\partial t}e_n\right)u(x, t) = 0,$$

can be written as

$$\overline{\partial} u = F(x,t), \tag{2.22}$$

$$\left(\overline{\partial} + \frac{\partial}{\partial t} e_n\right) F = 0. \tag{2.23}$$

The corresponding problems for (2.22) and (2.23) are correctly posed and solvable in quadratures.

B&IVP for the nonhomogeneous equations corresponding to the above homogeneous equations will be solved too. In this case the boundary-initial conditions can be supposed homogeneous. We will consider only one of them; others can be solved in the same way.

**Problem.** Find the solution of the equation

$$\Delta\left(\Delta - \frac{\partial^2}{\partial t^2}\right) u(x,t) = F(x,t), \tag{2.24}$$

$$x(x_0, \ldots, x_{n-1}) \in R^n, \quad t > 0, \quad x_{n-1} > 0,$$

that vanishes at infinity by the conditions

$$u(x,0) = 0, \quad \frac{\partial u}{\partial t} = 0, \quad t = 0, \tag{2.25}$$

$$u(x,t) = 0, \quad x_{n-1} = 0, \quad t > 0. \tag{2.26}$$

**Solution.** Let

$$\Delta u = F_1(x,t), \tag{2.27}$$

$$\Delta F_1 - \frac{\partial^2 F_1}{\partial t^2} = F(x,t). \tag{2.28}$$

Then by force of (2.25) for $F_1(x,t)$, one has Cauchy homogeneous conditions for the nonhomogeneous wave equation (2.28). Thus it is defined in quadratures. Hence $u(x,t)$ is defined as the solution of (2.27) with the condition (2.26).

**Lorentz Transformation.** As is well known, the Lorentz transformation has very important applications in relativity theory. In [Ob1] they were constructed using Clifford analysis. In particular, the equation

$$\overline{\partial} u = 0 \quad \text{in} \quad R_{(n,n-1)} \quad x(x_0, \ldots, x_n), \tag{2.29}$$

is invariant with respect to linear transformations of $x$ and the basis $e_k$ ($k = 0, 1, \ldots, n$)

$$y_1 = p(x_1 + vx_n), \quad y_j = x_j, \quad j = 0, 2, \ldots, n-1, \quad y_n = p(x_n + vx_1),$$
$$e_1 = p(f_1 + vf_n), \quad e_j = f_j, \quad j = 0, 2, \ldots, n-1, \quad e_n = p(f_n + vf_1), \quad (2.30)$$

where $p = \dfrac{1}{\sqrt{1-v^2}}$ is a scalar constant, $|v| < 1$. It is easy to see that

$$e_1 e_n = f_1 f_n. \quad (2.31)$$

If $f_A$ is an arbitrary basis of $R_{(n,n-1)}$ and if $A(\alpha_1, \ldots, \alpha_k)$ contains either both 1 and $n$ or neither 1 nor $n$, then one has $e_A = f_A$.

For the other basis elements we have

$$e_{1A} = \frac{1}{p}\left(f_{1A} + (-1)^k v f_{An}\right),$$
$$e_{An} = \frac{1}{p}\left(f_{An} + (-1)^k v f_{1A}\right). \quad (2.32)$$

Note that if (2.29) is represented in terms of the basis, then the coefficients are not individually invariant with respect to the above transformations. The coefficients of the $e_A$, which are invariant, are invariant themselves. The other coefficients, i.e., the coefficients of the noninvariant basis elements, are not invariant, but the sum of the addends containing $e_{An}$ and $e_{1A}$ are invariant. The Maxwell and Dirac equations are invariant with respect to the Lorentz transformations in the sense described above.

Now the question is, is the pluriregular equation (2.1) invariant with respect to the Lorentz transformations in above sense? The answer is affirmative because, in the case of $m = 2$, equation (2.1) can be written as

$$\overline{\partial} u = F,$$
$$\overline{\partial} F = 0,$$

and these equations are invariants.

As is well known, for the wave equation and for the corresponding simple hyperbolic systems with one space variable and one time variable like Cauchy's problem, some characteristic problems such as the Goursat and Darboux problems can be solved explicitly in a simple way using a general representation of the solution. But when there is more than one space variable, it is not always possible to solve the characteristic problems explicitly. So for the pluriregular equation (2.1), characteristic problems in multidimensional spaces are very complicated. That is why we will consider (2.1) in the space $R_{(1,0)}$

$$u(x) = u_0 e_0 + u_1 e_1, \quad x = x_0 e_0 + x_1 e_1, \quad e_1^2 = e_0,$$
$$\overline{\partial} = \frac{\partial}{\partial x_0} e_0 + \frac{\partial}{\partial x_1} e_1. \quad (2.33)$$

## 2. B&IVP for ... and Harmonic Klein–Gordon Equations

**Goursat problem.** Let $m = 2$. If $m > 2$, the problem is solved gradually. Then (2.1) can be written as

$$\frac{\partial u_0}{\partial x_0} + \frac{\partial u_1}{\partial x_1} = F_0(x_0, x_1),$$
$$\frac{\partial u_0}{\partial x_1} + \frac{\partial u_1}{\partial x_0} = F_1(x_0, x_1), \quad (2.34)$$

where $F_0, F_1$ are the solutions of the equations

$$\frac{\partial F_0}{\partial x_0} + \frac{\partial F_1}{\partial x_1} = 0,$$
$$\frac{\partial F_0}{\partial x_1} + \frac{\partial F_1}{\partial x_0} = 0. \quad (2.35)$$

Find the regular solution of (2.34), (2.35) for $x \in R^2$ by the conditions along the characteristics

$$u_0(x_0, x_0) = \varphi(x_0), \quad u_1(x_0, -x_0) = \psi(x_0), \quad (2.36)$$
$$F_0(x_0, x_0) = \varphi_1(x_0), \quad F_1(x_0, -x_0) = \psi_1(x_0). \quad (2.37)$$

**Solution.** By conditions (2.37), the solution of (2.35) can be represented as

$$F_0 = \varphi_1\left(\frac{x_0 + x_1}{2}\right) + \psi_1\left(\frac{x_0 - x_1}{2}\right) - \psi_1(0),$$
$$F_1 = -\varphi_1\left(\frac{x_0 + x_1}{2}\right) + \psi_1\left(\frac{x_0 - x_1}{2}\right) + \varphi_1(0). \quad (2.38)$$

Then the solution of equation (2.34) by conditions (2.36) can be defined as

$$u_0 = \varphi\left(\frac{x_0 + x_1}{2}\right) + \psi\left(\frac{x_0 - x_1}{2}\right) - \psi(0) + \frac{x_0 - x_1}{2}\varphi_1\left(\frac{x_0 + x_1}{2}\right) +$$
$$+ \frac{x_0 + x_1}{2}\psi_1\left(\frac{x_0 - x_1}{2}\right) - x_0\psi_1(0),$$
$$u_1 = -\varphi\left(\frac{x_0 + x_1}{2}\right) + \psi\left(\frac{x_0 - x_1}{2}\right) + \varphi(0) - \frac{x_0 - x_1}{2}\varphi_1\left(\frac{x_0 + x_1}{2}\right) +$$
$$+ \frac{x_0 + x_1}{2}\psi_1\left(\frac{x_0 - x_1}{2}\right) + \varphi_1(0)x_0.$$

It is easy to see that it satisfies all conditions if the given functions have first-order continuous derivatives.

Replacing conditions (2.36), (2.37) with the conditions

$$u_0(x_0, x_0) = \varphi(x_0), \quad u_0(x_0, -x_0) = \psi(x_0), \quad (2.39)$$

$$F_0(x_0, x_0) = \varphi_1(x_0), \quad F_0(x_0, -x_0) = \psi_1(x_0), \tag{2.40}$$

where the given functions must satisfy the compatibility conditions

$$\varphi(0) = \psi(0), \quad \varphi_1(0) = \psi_1(0). \tag{2.41}$$

This solution can be represented explicitly too.

## 3  Pluri-Beltrami Hyperbolic Equations

First consider the Beltrami equation and its generalization in $R_{(1,0)}$, i.e., with the variable $x = x_0 e_0 + x_1 e_1$:

$$\overline{\partial} u + q \partial u = 0, \tag{3.1}$$
$$\overline{\partial} u + q_1 \partial u + q_2 \overline{\partial} \overline{u} = 0. \tag{3.2}$$

As was obtained in the elliptic case, one has [Ob2]:
(1) (3.1) is hyperbolic with the condition $|q| \neq 1$; moreover, the case $|q| > 1$ can be reduced to $|q| < 1$. (3.3)
(2) (3.2) is hyperbolic if

$$|q_1| + |q_2| < 1 \text{ and } \big||q_1| - |q_2|\big| < 1, \text{ i.e., equivalently}$$
$$|q_1| + |q_2| < 1, \tag{3.4}$$

or

$$|q_1| + |q_2| > 1 \text{ and } \big||q_1| - |q_2|\big| > 1, \text{ i.e., more simply } \big||q_1| - |q_2|\big| > 1, \tag{3.5}$$

where (3.5) can be reduced to (3.4).
If

$$\big||q_1| - |q_2|\big| < 1 \text{ and } |q_1| + |q_2| > 1, \tag{3.6}$$

then (3.2) is elliptic.

As in the elliptic case, the equation (3.2) with vectorial $q_1, q_2 \in R_{(n,n-1)}$ is hyperbolic with the above conditions (3.4), (3.5) and elliptic with (3.6).
Let $q_1, q_2 \in R_{(1,0)}$. Because $e_1^2 = e_0$, $|q_1|^2 = q_{11}^2 - q_{12}^2$, one has

$$q_1 = |q_1| \left( \frac{\sqrt{q_{11} + |q_1|}}{\sqrt{2}\sqrt{|q_1|}} e_0 + \frac{q_{12} e_1}{\sqrt{2}\sqrt{|q_1|}\sqrt{q_{11} + |q_1|}} \right)^2 \equiv |q_1|(a_0 e_0 + a_1 e_1)^2,$$

## 3. Pluri-Beltrami Hyperbolic Equations

$$q_2 \equiv |q_2|(b_0 e_0 + b_1 e_1)^2.$$

In the hyperbolic case, any $q_{11}, q_{12}$ with $q_{11}^2 - q_{12}^2 = 1$ can be represented with the help of hyperbolic sine and cosine

$$q_{11} = \operatorname{ch} \alpha, \quad q_{12} = \operatorname{sh} \alpha, \quad q = e_0 \operatorname{ch} \alpha + e_1 \operatorname{sh} \alpha,$$

so that

$$q^2 = (e_0 \operatorname{ch} \alpha + e_1 \operatorname{sh} \alpha)^2 = (\operatorname{ch}^2 \alpha + \operatorname{sh}^2 \alpha) e_0 + 2 e_1 \operatorname{sh} \alpha \operatorname{ch} \alpha = e_0 \operatorname{ch} 2\alpha + e_1 \operatorname{sh} 2\alpha.$$

Analogously, in the $n$-dimensional case, if $e_0^2 = e_0, e_1^2 = \cdots = e_{n-1}^2 = -e_0, e_n^2 = e_0$, then

$$q = e_0 \cos \alpha_0 + e_1 \sin \alpha_0 \sin \alpha_1 + e_2 \sin \alpha_0 \sin \alpha_1 \cos \alpha_2 + \cdots +$$
$$+ e_{n-1} \sin \alpha_0 \sin \alpha_1 \cdots \operatorname{ch} \alpha_{n-1} + e_n \sin \alpha_0 \sin \alpha_1 \cdots \operatorname{sh} \alpha_{n-1}.$$

In $R_{(2,1)}$, for instance, we have

$$q = e_0 \cos \alpha_0 + e_1 \sin \alpha_0 \operatorname{ch} \alpha_1 + e_2 \sin \alpha_0 \operatorname{sh} \alpha_1, \quad |q| = 1,$$
$$q^2 = e_0 \cos 2\alpha_0 + e_1 \sin 2\alpha_0 \operatorname{ch} \alpha_1 + e_2 \sin 2\alpha_0 \operatorname{sh} \alpha_1,$$

and for any $m \geq 1$, one has the De Moivre-like formula

$$q^m = e_0 \cos m\alpha_0 + e_1 \sin m\alpha_0 \operatorname{ch} \alpha_1 + e_2 \sin m\alpha_0 \operatorname{sh} \alpha_1.$$

With the representation for $q_1, q_2$, equation (3.2) in $R_{(1,0)}$ can be written as

$$\bar{\partial}_y v + |q_1| \partial_y v + |q_2| \bar{\partial}_y \bar{v} = 0, \tag{3.7}$$

where

$$v = (b_0 e_0 + b_1 e_1)^{-1} u, \quad y = (a_0 e_0 - a_1 e_1) x. \tag{3.8}$$

Hence if $v = v_0 e_0 + v_1 e_1$, $|q_1| \equiv q_1$, $|q_2| \equiv q_2$, then equation (3.7) can be written as

$$\frac{\partial v_0}{\partial y_0}(1+q_1+q_2) + (1-q_1-q_2)\frac{\partial v_1}{\partial y_1} = 0,$$
$$\frac{\partial v_0}{\partial y_1}(1-q_1+q_2) + (1+q_1-q_2)\frac{\partial v_1}{\partial y_0} = 0. \qquad (3.9)$$

Then it is easy to obtain that $v_0, v_1$ is the solution of the equation

$$\frac{\partial^2 v_0}{\partial y_0^2}[(1+q_1)^2 - q_2^2] - \frac{\partial^2 v_0}{\partial y_1^2}[(1-q_1)^2 - q_2^2] = 0.$$

Hence by the transformation

$$\xi = [(1-q_1)^2 - q_2^2]^{1/2} y_0, \quad \eta = [(1+q_1)^2 - q_2^2]^{1/2} y_1, \qquad (3.10)$$

one can see that $v_0, v_1$ as functions of $\xi, \eta$ are solutions of the wave equation. System (3.9) by (3.10) can be written in the form

$$\frac{\partial v_0}{\partial \xi}[(1+q_2)^2 - q_1^2]^{1/2} + \frac{\partial v_1}{\partial \eta}[(1-q_2)^2 - q_1^2]^{1/2} = 0,$$
$$\frac{\partial v_0}{\partial \eta}[(1+q_2)^2 - q_1^2]^{1/2} + \frac{\partial v_1}{\partial \xi}[(1-q_2)^2 - q_1^2]^{1/2} = 0. \qquad (3.11)$$

Consider the functions

$$w_0 = v_0[(1+q_2)^2 - q_1^2]^{1/2},$$
$$w_1 = v_1[(1-q_2)^2 - q_1^2]^{1/2}. \qquad (3.12)$$

Thus equation (3.2) with real coefficients by the transformation of variables (3.10) and of functions (3.12) is reduced to the form

$$\bar{\partial} w = 0, \qquad (3.13)$$

where

$$w = w_0 e_0 + w_1 e_1, \quad \bar{\partial} = \frac{\partial}{\partial \xi} e_0 + \frac{\partial}{\partial \eta} e_1.$$

Hence $w(\zeta)$, $\zeta = \xi e_0 + \eta e_1$, is a solution of the wave equation too. The IVP that are solved for (3.13) can be solved for (3.2) explicitly in quadratures.

## 3. Pluri-Beltrami Hyperbolic Equations

Now consider the pluri-Beltrami equation in the space $R_{(1,0)}$

$$(\overline{\partial}_x + q\partial_x)^m u = 0, \quad m \geq 1, \quad q = const. \quad (3.14)$$

Since by the transformation

$$y = (a_0 e_0 - a_1 e_1)x, \quad q = |q|(a_0 e_0 + a_1 e_1)^2, \quad \xi = y - |q|\overline{y},$$

this equation can be reduced to the form (2.1), all IVP solved for it can be solved for equation (3.14) too.

Consider equation (3.1) in $R_{(n,n-1)}$. We can prove that, by a uniquely defined linear transformation of independent variables, this equation can be reduced to the regular equation $\overline{\partial}_y u = 0$ by analogy with the elliptic case. Correspondingly, equation (3.14) in $R_{(n,n-1)}$ can be reduced to the pluriregular equation $\overline{\partial}^m u = 0$. Therefore, all IVP that are solved for this equation can be solved for the pluri-Beltrami equation too.

As we know, the space $R_{(3,2)}$ is remarkable because the equation $\overline{\partial}_x u = 0$ in this space represents the Dirac equation of relativistic quantum mechanics or Maxwell's equations. Therefore, equation (3.1) in this space can be called the Beltrami–Dirac equation.

Equation (3.2) considered in $R_{(n,n-1)}$ cannot be reduced to a regular equation. But if in place of $\overline{u}$ we write $\widetilde{u}$, i.e.,

$$\overline{\partial}u + q_1 \partial u + q_2 \overline{\partial}\widetilde{u} = 0, \quad (3.15)$$

then using the transformation defined by analogy with the elliptic case, this equation can be reduced to (3.15) with real $q_2$. Then, supposing $q_1$ is real, we can reduce this equation to a regular equation by transformation of the independent variables and unknown functions.

Now consider the equation

$$\overline{\partial}_x u + q\partial_x u + \widetilde{u}h = 0, \quad |q| < 1, \quad (3.16)$$

where $q$ and $h$ are vectorial constants of the space $R_{(n,n-1)}$. By the uniquely defined linear transformation of independent variables

$$y_k = \sum_{i=0}^{n} a_{ki} x_i, \quad k = 0, 1, \ldots, n,$$

equation (3.16) can be reduced to

III. Hyperbolic and Plurihyperbolic Equations in Clifford Analysis

$$\bar{\partial}_y u + \tilde{u} h = 0. \tag{3.17}$$

One can see that if $u$ solves (3.17), then it also solves the equation

$$\partial \bar{\partial} u - |h|^2 u = 0, \tag{3.18}$$

which in $R_{(n,n-1)}$ is the Klein–Gordon equation. Thus the solution of (3.16) can be represented as

$$u(x) = \varphi(y_0, y_1, \ldots, y_n), \tag{3.19}$$

where $\varphi(y)$ is a solution of (3.17).

All IVP that are solved for equation (3.17) can be solved for equation (3.16) using representation (3.19).

# IV

# Parabolic and Pluriparabolic Equations in Clifford Analysis

## 0 Introduction

Systems of parabolic equations that are related to the heat and polyheat equations will be considered. First consider the heat equation

$$\Delta u = \frac{\partial u}{\partial t}, \quad u(x,t), \quad x_n \equiv t, \quad t > 0, \quad x \in R^n, \quad n \geq 1, \tag{1}$$

where $\Delta$ is the Laplace operator with respect to the variables $x(x_0, \ldots, x_{n-1})$.

**Cauchy problem.** Find the solution of (1) with the condition

$$u(x, 0) = \varphi(x). \tag{2}$$

**Solution.** The FIT with respect to $x_0, \ldots, x_{n-1}$ of equation (1) gives us

$$\frac{d\widehat{u}(y,t)}{dt} + |y|^2 \widehat{u} = 0, \quad y(y_0, \ldots, y_{n-1}) \in R^n,$$

with the condition $\widehat{u}(y, 0) = \widehat{\varphi}(y)$. Thus it is defined as

$$\widehat{u}(y, t) = \widehat{\varphi}(y) e^{-|y|^2 t}.$$

This is a function of the class $L(R^n)$, $\forall t > 0$, and we can use the inversion theorem

$$u(x,t) = \frac{1}{(\sqrt{2\pi})^n} \int_{R^n} \widehat{\varphi}(y) e^{-|y|^2 t} e^{i(x,y)} dy,$$

which by force of the properties of convolution and with the Gauss–Weierstrass kernel can be represented in the form

$$u(x,t) = \frac{1}{(2\sqrt{\pi t})^n} \int_{R^n} \varphi(y) \exp\left[-\frac{|x-y|^2}{4t}\right] dy, \qquad (3)$$

known as Poisson's formula.

The function

$$g(x,t) = \frac{1}{(\sqrt{t})^n} \exp\left[-\frac{|x|^2}{4t}\right],$$

with the singularity at the point $x = 0$, $t = 0$, is a fundamental solution of equation (1). By force of (3) one can easily prove

$$\int_{R^n} u(x,t)\,dx = \int_{R^n} \varphi(x)\,dx,$$

which states that in the space of $x \in R^3$, the quantity of heat does not change in time.

Now Cauchy's problem for the nonhomogeneous equation

$$\frac{\partial u}{\partial t} = \Delta u + f(x,t), \qquad (4)$$

can be solved easily. In fact, if $v(x,t,\tau)$ is the solution of the homogeneous equation (1) for $t > \tau$, $x \in R^n$ with the condition

$$v(x,\tau,\tau) = f(x,\tau),$$

then the solution of (4) with the condition

$$u(x,0) = 0,$$

can be defined as

$$u(x,t) = \int_0^t v(x,t,\tau)d\tau,$$

i.e., by force of (3), it is represented as

$$u(x,t) = \frac{1}{(2\sqrt{\pi})^n} \int_0^t \frac{d\tau}{(t-\tau)^{n/2}} \int_{R^n} f(y,\tau) \exp\left[-\frac{|x-y|^2}{4(t-\tau)}\right] dy. \quad (5)$$

Consider the Clifford algebras $R^0_{(n)}$ ($n \geq 1$) with the rules

$$e_0^2 = e_0, \quad e_k^2 = -e_0, \quad k = 1, \ldots, n-1, \quad e_n^2 = 0, \\ e_j e_k + e_k e_j = 0, \quad j \neq k. \quad (6)$$

Consider the equation [Ob1]

$$\bar{\partial} u - P_n u = 0, \quad u(x) = u(x_0, x_1, \ldots, x_n), \quad x_n \equiv t, \quad (7)$$

where the linear operator $P_n$ is defined by the condition

$$\partial P_n u = \frac{\partial u}{\partial t}, \quad (8)$$

i.e., the solution of (7) is the solution of the heat equation (1) too. The solution $u(x)$ can be represented in the form

$$u(x) = \sum_{A \neq n} u_A e_A + \sum_{A \neq n} u_{An} e_A e_n, \quad A(\alpha_1, \ldots, \alpha_k). \quad (9)$$

**Theorem 1.** *Let $u(x) \in R^0_{(n)}$ be the solution of equation (7), where*

$$P_n u = -\sum_{A \neq n} (-1)^k u_{An} e_A. \quad (10)$$

*Then condition (8) is true. Moreover, by condition (8), $P_n u$ is defined uniquely.*

By force of (9)

$$\bar{\partial} u = \sum_{A \neq n} (\bar{\partial} u_A) e_A + \sum_{A \neq n} (\bar{\partial} u_{An}) e_A e_n. \quad (11)$$

Because $P_n u$ does not contain a term with $e_n$, from equation (7) it follows that in $\bar{\partial} u$ the addends with $e_n$ must be zero. Thus by (6), (11) one can get

$$\sum_{A \neq n} \frac{\partial u_A}{\partial x_n} e_n e_A + \sum_{A \neq n} (-1)^k \frac{\partial u_{An}}{\partial x_0} e_n e_A - \sum_{\substack{A \neq n \\ j=1}}^{n-1} (-1)^k \frac{\partial u_{An}}{\partial x_j} e_n e_j e_A = 0. \quad (12)$$

By virtue of (10), we can obtain

$$\partial P_n u = - \sum_{A \neq n} (-1)^k \frac{\partial u_{An}}{\partial x_0} e_A + \sum_{\substack{A \neq n \\ j=1}}^{n-1} (-1)^k \frac{\partial u_{An}}{\partial x_j} e_j e_A + \sum_{A \neq n} \frac{\partial u_{An}}{\partial x_n} e_A e_n.$$

Applying (9), (12) gives equality (8). So if $u(x)$ is a solution of (7), then it is also a solution of the heat equation (1). Now it is easy to show that $P_n$ with property (8) for $x(x_0, \ldots, x_{n-1}), t > 0$, that vanishes at infinity is defined uniquely. Let a second operator $Q_n u$ exist that satisfies (8), i.e.,

$$\partial Q_n u = \frac{\partial u}{\partial t}.$$

Then by force of (8), we get

$$\Delta [P_n u - Q_n u] = 0,$$
$$\frac{\partial P_n u}{\partial t} = \frac{\partial Q_n u}{\partial t}.$$

By the Liouville theorem, $P_n u - Q_n u$, a harmonic function in $R^n$ that vanishes at infinity, is zero; i.e., in $R_{(n)}^0$ equation (7) is the only one related to the heat equation.

Equation (7) with (9), (10) can be written as

$$\sum_{\substack{A \neq n \\ j=0}}^{n-1} \frac{\partial u_A}{\partial x_j} e_j e_A + \sum_{A \neq n} (-1)^k u_{An} e_A = 0, \quad (11_1)$$

$$\sum_{A \neq n} \frac{\partial u_A}{\partial x_n} e_A + m_{A \neq n} (-1)^k \frac{\partial u_{An}}{\partial x_0} e_A - \sum_{\substack{j=1 \\ A \neq n}}^{n-1} (-1)^k \frac{\partial u_{An}}{\partial x_j} e_j e_A = 0. \quad (12_1)$$

**Theorem 2.** *Let $u$ satisfy $(11_1)$ and each of $u_A$ ($A \neq n$) be a solution of the heat equation (1). Then $u$ is a solution of $(12_1)$ too.*

## 0. Introduction

If (11₁) is differentiated with respect to $x_0$ and to $x_i$, one can get respectively

$$\sum_{\substack{A\neq n \\ j=0}}^{n-1} \frac{\partial^2 u_A}{\partial x_0 \partial x_j} e_j e_A + \sum_{A\neq n}(-1)^k \frac{\partial u_{An}}{\partial x_0} e_A = 0, \tag{13}$$

$$\sum_{i=1}^{n-1}\sum_{\substack{A\neq n \\ j=0}}^{n-1} \frac{\partial^2 u_A}{\partial x_i \partial x_j} e_i e_j e_A + \sum_{A\neq n}\sum_{i=1}^{n-1}(-1)^k \frac{\partial u_{An}}{\partial x_i} e_i e_A = 0. \tag{14}$$

One can check that

$$\sum_{j=0}^{n-1} \frac{\partial^2 u_A}{\partial x_0 \partial x_j} e_j - \sum_{i=1}^{n-1}\sum_{j=0}^{n-1} \frac{\partial^2 u_A}{\partial x_j \partial x_i} e_i e_j = \Delta u_A = \frac{\partial u_A}{\partial t}.$$

So by taking the difference of equations (13), (14), one can obtain equation (12₁).

This theorem will be used to solve certain IVP. Consider particular cases of equation (7).

Let $x(x_0, x_1, x_2)$, $x_2 \equiv t$, $u(x) \in R_{(2)}^0$,

$$u(x) = u_0 e_0 - u_1 e_1 - u_2 e_2 - u_{12} e_1 e_2.$$

By (10) we have

$$P_2 u = u_2 e_0 - u_{12} e_1,$$

and equation (7) is equivalent to the parabolic system

$$\frac{\partial u_0}{\partial x_0} + \frac{\partial u_1}{\partial x_1} - u_2 = 0, \quad \frac{\partial u_{12}}{\partial x_0} + \frac{\partial u_2}{\partial x_1} - \frac{\partial u_1}{\partial t} = 0,$$
$$\frac{\partial u_0}{\partial x_1} - \frac{\partial u_1}{\partial x_0} + u_{12} = 0, \quad \frac{\partial u_{12}}{\partial x_1} - \frac{\partial u_2}{\partial x_0} + \frac{\partial u_0}{\partial t} = 0. \tag{15}$$

This can be considered the parabolic analogue of the Moisil–Theodorescu system.

Consider the complex functions

$$w_1 = u_0 - i u_1, \quad w_2 = u_2 + i u_{12}.$$

Then (15) can be written in the complex form

$$2\frac{\partial w_1}{\partial \bar{z}} - \bar{w}_2 = 0, \quad 2\frac{\partial w_2}{\partial \bar{z}} - \frac{\partial \bar{w}_1}{\partial t} = 0, \quad z = x_0 + ix_1. \tag{16}$$

Therefore,

$$\Delta w_k - 2\frac{\partial w_k}{\partial t} = 0, \quad k = 1, 2.$$

Taking the previous theorem into consideration, it is sufficient to consider, for instance the first equation of (16) and the heat equation for $w_1$ in order to define $w_1, w_2$ as the solution to (16).

Now let $x(x_0, x_1, x_2, x_3)$, $x_3 \equiv t$, $u \in R^0_{(3)}$, and

$$u(x) = u_0 e_0 - \sum_{1}^{3} u_k e_k - u_{12} e_1 e_2 - u_{13} e_1 e_3 - u_{23} e_2 e_3 - u_{123} e_1 e_2 e_3. \tag{17}$$

By (10) in this case

$$P_3 u = u_3 e_0 - u_{13} e_1 - u_{23} e_2 + u_{123} e_1 e_2. \tag{18}$$

Then equation (7) is equivalent to the system

$$\begin{aligned} \text{div } U - \varphi = 0, \quad & \text{div } V + \frac{\partial \psi}{\partial t} = 0, \\ \text{grad } \psi + \text{rot } U + V = 0, \quad & \text{grad } \varphi + \text{rot } V - \frac{\partial U}{\partial t} = 0, \end{aligned} \tag{19}$$

where $U \equiv (u_0, u_1, u_2)$, $V \equiv (u_{123}, u_{23}, -u_{13})$, $u_{12} \equiv \varphi$, $u_3 = \psi$, and the operators grad, div, rot are taken with respect to $x_0, x_1, x_2$. This system can be considered the parabolic analogue of the Dirac equations of relativistic quantum mechanics.

## 1   IVP for Parabolic Systems in Clifford Analysis

**Cauchy's problem.** Find the regular solution of (7) for $x(x_0, \ldots, x_{n-1}) \in R^n$, $t > 0$ with $2^{n-1}$ initial conditions

$$u_A(x, 0) = \varphi_A(x), \quad A(\alpha_1, \ldots, \alpha_k) \in (0, 1, \ldots, n-1). \tag{1.1}$$

**Solution.** Because each $u_A$ is also the solution of the heat equation by the conditions (1.1), all $u_A$ ($A \neq n$) are defined by (3). The remaining $2^{n-1}$ unknowns $u_{An}$ ($A \neq n$) are defined from equation ($11_1$). Then $u_A, u_{An}$ satisfy ($12_1$) too.

If in place of conditions (1.1) we have

$$u_{An}(x,0) = \varphi_A(x), \quad A(\alpha_1,\ldots,\alpha_k) \in \{0,1,\ldots,n-1\}, \quad \text{for } x \in R^n, \quad (1.2)$$

then $u_{An}(x,t), t > 0, (A \neq n)$ are represented by (3). By equation $(12_1)$ the derivatives $\dfrac{\partial u_A}{\partial t}$ are defined for each $A \neq n$; i.e., all $u_A$ $(A \neq n)$ can be represented as

$$u_A(x,t) = f_A(x,t) + u'_A(x), \quad A \neq n,$$

where the $f_A(x,t)$ are known and the $u'_A(x)$, $x \in R^n$ must be defined by equation $(11_1)$. Substituting $u_A$ and $u_{An}$ into equation $(11_1)$, one obtains for $u'_A(x)$, $x \in R^n$ an elliptic equation in $R_{(n-1)}$

$$\overline{\partial} u' = 0, \quad u' = \sum_{A \neq n} u'_A e_A. \quad (1.3)$$

Thus $u'(x)$ with values in $R_{(n-1)}$ is the solution of the regular elliptic equation (1.3) in $R^n$. So by Liouville's theorem, it is zero. Hence $u_A, u_{An}$ are defined by (1.2) uniquely.
Let $u(x,t)$ be the solution of the nonhomogeneous equation in $R^0_{(n)}$

$$\overline{\partial} u - P_n u = f(x,t), \quad (1.4)$$

where

$$f(x,t) = \sum_{A \neq n} f_A(x,t) e_A, \quad (1.5)$$

or

$$f(x,t) = \sum_{A \neq n} f_{An}(x,t) e_A e_n. \quad (1.6)$$

In this case $u(x,t)$ is also the solution of the nonhomogeneous heat equation

$$\Delta_{(n)} u - \frac{\partial u}{\partial t} = \partial f,$$
$$\partial f = \sum F_A e_A. \quad (1.7)$$

Just as equation (7) was written in the form $(11_1)$, $(12_1)$, equation (1.4) with (1.5) can be written as $(11_1)$, $(12_1)$, where the right-hand side of $(11_1)$ is (1.5). But (1.4)

158    IV. Parabolic and Pluriparabolic Equations in Clifford Analysis

with (1.6) can be written as $(11_1)$, $(12_1)$, where the right-hand side of $(12_1)$ is defined by (1.6). The theorem formulated above after $(12_1)$ for the equation (1.4) follows.

Let $u(x, t)$ satisfy $(11_1)$ with the right-hand side (1.5) and each $u_A$ $(A \neq n)$ be the solution of equation (1.7). Then $u(x, t)$ is the solution of $(12_1)$, i.e., the solution of (1.4).

**Problem.** Find a regular solution of (1.4) subject to the $2^{n-1}$ conditions

$$u_A(x, 0) = 0, \quad A \neq n, \quad x(x_0, \ldots, x_{n-1}). \tag{1.8}$$

Since by (1.7) the $u_A$ are the solutions of the equation

$$\Delta_{(n)} u_A - \frac{\partial u_A}{\partial t} = F_A, \quad A \neq n,$$

by condition (1.8), they are represented by (5), where $f(x, t)$ is the right-hand side of the last equation. Then the remaining unknown $u_{An}$ is defined by equation $(11_1)$ with the right-hand side (1.5). Hence, by the above theorem, the solution of the problem is defined completely.

Consider the problem for the half space $x_{n-1} \geq 0$. Find the regular solution of (1) for $x_{n-1} > 0, t > 0$ with the conditions

$$u(x, 0) = \psi(x), \quad u(x_0, \ldots, x_{n-2}, 0, t) = 0, \quad t > 0, \quad x'(x_0, \ldots, x_{n-2}) \in R^{n-1}. \tag{1.9}$$

The function $\psi(x)$ must satisfy the compatibility condition

$$\psi(x_0, \ldots, x_{n-2}, 0) = 0.$$

The solution of this problem can be easily reduced to the problem (2) in $R^n, t > 0$. In fact, define the function

$$\varphi(x) = \begin{cases} \psi(x), & \text{for } x_{n-1} \geq 0, \\ -\psi(x_0, \ldots, x_{n-2}, -x_{n-1}), & \text{for } x_{n-1} \leq 0. \end{cases} \tag{1.10}$$

Then (3), where $\varphi(x)$ is an odd function defined by (1.10), is the solution of the problem.

If we consider the problem with the conditions

$$u(x,0) = \psi(x), \quad x_{n-1} \geq 0, \quad \frac{\partial u(x,t)}{\partial x_{n-1}} = 0, \quad \text{for } x_{n-1} = 0, \qquad (1.11)$$

the function $\psi(x)$ must be continued in even form for $x_{n-1} < 0$. Its solution is represented by (3).

## 2 Pluriparabolic Systems and Polyheat Equations

Let $u(x,t) \in R^0_{(n)}$ and consider the high-order equations

$$(\bar{\partial} - P_n)^m u = 0, \quad m \geq 1. \qquad (2.1)$$

By force of (7), (8), $u(x,t)$ is also the solution of the polyheat equation

$$\left(\Delta - \frac{\partial}{\partial t}\right)^m u = 0, \quad x(x_0, \ldots, x_{n-1}). \qquad (2.2)$$

In the case $m = 2$, (2.2) is called the biheat equation. Cauchy's problem for (2.2) is as follows. Find the solution $u(x,t)$ of (2.2), $t > 0$ by the conditions

$$u(x,0) = \varphi_0(x), \quad \frac{\partial u}{\partial t} = \varphi_1(x), \ldots, \frac{\partial^{m-1} u}{\partial t^{m-1}} = \varphi_{m-1}(x), \quad t = 0. \qquad (2.3)$$

First solve this problem for $m = 2$, i.e., for the equation

$$\left(\Delta - \frac{\partial}{\partial t}\right)^2 u = 0, \qquad (2.4)$$

with

$$u(x,0) = \varphi_0(x), \quad \frac{\partial u}{\partial t} = \varphi_1(x), \quad t = 0. \qquad (2.5)$$

**Solution.** From (2.4) it follows that

$$\Delta u - \frac{\partial u}{\partial t} = F(x,t), \qquad (2.6)$$

$$\Delta F - \frac{\partial F}{\partial t} = 0, \qquad (2.7)$$

160    IV. Parabolic and Pluriparabolic Equations in Clifford Analysis

where $F$ is defined for $t = 0$ by (2.5), (2.6). Thus $F(x, t)$ is represented as (3). Then the solution of (2.6) is represented as

$$u(x, t) = u_1(x, t) + u_2(x, t), \tag{2.8}$$

where $u_1, u_2$ are solutions of the equations

$$\Delta u_1 - \frac{\partial u_1}{\partial t} = 0, \quad \text{with the condition} \quad u_1(x, 0) = \varphi_0(x), \tag{2.9}$$

$$\Delta u_2 - \frac{\partial u_2}{\partial t} = F(x, t), \quad \text{with the condition} \quad u_2(x, 0) = 0. \tag{2.10}$$

Thus $u_1(x, t)$ is represented as (3), and $u_2$ is defined by (5). Then for any $m \geq 2$, the problem can be solved by the inductive method. But the solution can be obtained directly using FIT. From (2.2), (2.3) it follows that the FIT of $u(x, t)$ with respect to $x_0, \ldots, x_{n-1}$ satisfies

$$\left(-\frac{d}{dt} - |\tau|^2\right)^m \widehat{u} = 0, \quad \tau(\tau_0, \tau_1, \ldots, \tau_{n-1}),$$
$$\frac{d^k \widehat{u}(\tau, t)}{dt^k} = \widehat{\varphi}_k(\tau), \quad t = 0, \quad k = 0, \ldots, m-1. \tag{2.11}$$

Thus $\widehat{u}$ is represented as

$$\widehat{u}(\tau, t) = (a_0 + a_1 t + \ldots + a_{m-1} t^{m-1}) e^{-|\tau|^2 t}, \tag{2.12}$$

where the coefficients $a_k$ are defined by $\widehat{\varphi}_k$:

$$a_0 = \widehat{\varphi}_0, \quad a_1 = \widehat{\varphi}_1 + |\tau|^2 \widehat{\varphi}_0, \quad 2a_2 = \widehat{\varphi}_2 + 2|\tau|^2 \widehat{\varphi}_1 + |\tau|^4 \widehat{\varphi}_0,$$
$$3! a_3 = \widehat{\varphi}_3 + 3|\tau|^2 \widehat{\varphi}_2 + 3|\tau|^4 \widehat{\varphi}_1 + |\tau|^6 \widehat{\varphi}_0,$$

and in this way we can define all $a_k$. For example, in the case $m = 2$

$$\widehat{u}(\tau, t) = \left[\widetilde{\varphi}_0(\tau)(1 + |\tau|^2 t) + \widehat{\varphi}_1(\tau) t\right] e^{-|\tau|^2 t}, \quad t > 0.$$

Because it is a function of the class $L$ ($\tau \in R^n$), by the inverse theorem and (3), one can obtain

$$u(x, t) = \left(1 - t \frac{\partial}{\partial t}\right) Q_n(\varphi_0) + t Q_n(\varphi_1), \quad t > 0, \tag{2.13}$$

where

$$Q_n(\varphi) = \frac{1}{(2\sqrt{\pi t})^n} \int_{R^n} \varphi(y) \exp\left[-\frac{|x-y|^2}{4t}\right] dy. \qquad (2.14)$$

For any $m$, we can obtain the integral representation

$$u(x,t) = \sum_{p=0}^{m-1} \frac{t^p}{p!} \sum_{k=0}^{m-p-1} (-1)^k \frac{t^k}{k!} \frac{\partial^k}{\partial t^k} Q_n(\varphi_p). \qquad (2.15)$$

For equation (2.1) in the case $m = 2$, consider the following.

**Cauchy's problem.** Find the solution of (2.1) for $t > 0$ by the conditions

$$u_A(x, 0) = \varphi_A(x), \quad u_{An}(x, 0) = \psi_A(x), \quad A \in \{0, 1, \ldots, n-1\}; \qquad (2.16)$$

i.e., all components of $u$ are given.

**Solution.** Let

$$\bar{\partial} u - P_n u = F, \qquad (2.17)$$
$$\bar{\partial} F - P_n F = 0. \qquad (2.18)$$

In this case the right-hand side of $(11_1)$ is $F_A$ which, by force of (2.16), is defined for $t = 0$. Thus by force of (2.18), it is represented as the solution of (1.1) in quadratures. Then the solution of (2.17) by the first conditions of (2.16) is defined as for corresponding problems for the nonhomogeneous equation (1.4).

Note that if $u_0, \ldots, u_{m-1}$ are the solution of the heat equation then

$$u = \sum_{0}^{m-1} t^k u_k,$$

is the solution of the polyheat equation (2.2). This representation can also be successfully used to solve problem (2.3) in quadratures.

# 3 Parabolic Regular Equations of the Second Kind and IVP

Let $u(x) \in R^0_{(n,n-1)}$ $(e_1^2 = \ldots = e_{n-2}^2 = -e_0, e_{n-1}^2 = e_0, e_n^2 = 0)$ be the solution of equation (7), $n \geq 2$. The theorems of the previous section are also true in this case [Ob1].

**Theorem.** *If $P_n$ is defined by the equality (10), then conditions (8) are valid and $u(x)$ is also the solution of the equation*

$$\Delta_{n-1} u - \frac{\partial^2 u}{\partial x_{n-1}^2} - \frac{\partial u}{\partial t} = 0. \qquad (3.1)$$

**Theorem.** *Let $u(x)$ satisfy $(11_1)$ and each of $u_A$ ($A \neq n$) be the solution of equation (3.1). Then $u$ is also the solution of $(12_1)$, i.e., (7).*

The proofs of these theorems are quite analogous to those for $R_{(n)}^0$. Consider the particular cases for (7), which in our opinion are interesting.

(1) Let $u(x) \in R_{(2 \cdot 1)}^0$. Then $P_2 u = u_2 e_0 - u_{12} e_1$. Because $e_1^2 = e_0$, $e_2^2 = 0$, equation (7) is equivalent to the system

$$\frac{\partial u_0}{\partial x_0} - \frac{\partial u_1}{\partial x_1} - u_2 = 0, \quad \frac{\partial u_2}{\partial x_0} + \frac{\partial u_{12}}{\partial x_1} - \frac{\partial u_0}{\partial t} = 0,$$

$$\frac{\partial u_0}{\partial x_1} - \frac{\partial u_1}{\partial x_0} + u_{12} = 0, \quad \frac{\partial u_2}{\partial x_1} + \frac{\partial u_{12}}{\partial x_0} - \frac{\partial u_1}{\partial t} = 0, \quad t \equiv x_2. \qquad (3.2)$$

In this case each $u_k$ is also the solution of the equation

$$\frac{\partial^2 u}{\partial x_0^2} - \frac{\partial^2 u}{\partial x_1^2} = \frac{\partial u}{\partial t}. \qquad (3.3)$$

(2) Let $u(x) \in R_{(3,2)}^0$ and $u(x)$ be defined by (15). Then (7) is equivalent to the system (taking into consideration that $P_3 u$ has representation (16) and $e_1^2 = -e_0$, $e_2^2 = e_0$, $e_3^2 = 0$, $x(x_0, x_1, x_2)$)

$$\operatorname{div}_h U - \phi = 0, \quad \operatorname{div} V + \frac{\partial \psi}{\partial x_3} = 0,$$

$$\operatorname{grad}_h \psi + \operatorname{rot} U + V = 0, \quad \operatorname{grad} \phi + \operatorname{rot}_h V - \frac{\partial U}{\partial x_3} = 0, \qquad (3.4)$$

where $U(u_0, u_1, u_2)$, $V(u_{123}, u_{23}, -u_{13})$, $\phi = u_3$, $u_{12} = \psi$, and the operators $\operatorname{div}_h$, $\operatorname{grad}_h$, $\operatorname{rot}_h$, called hyperbolic operators, are obtained from the classical operators $\operatorname{div}$, $\operatorname{grad}$, $\operatorname{rot}$ with respect to the variables $x_0, x_1, x_2$, where $\frac{\partial}{\partial x_2}$ is replaced by $-\frac{\partial}{\partial x_2}$. In the classical case they can be called elliptic. One can easily check that

$$\operatorname{div}_h(\operatorname{grad} \phi) = \operatorname{div}(\operatorname{grad}_h \phi) = \frac{\partial^2 \phi}{\partial x_0^2} + \frac{\partial^2 \phi}{\partial x_1^2} - \frac{\partial^2 \phi}{\partial x_2^2},$$

$$\operatorname{div}_h(\operatorname{rot}_h U) = \operatorname{div} \operatorname{rot} U = 0, \qquad (3.5)$$

$$\operatorname{rot}_h(\operatorname{grad}_h \phi) = 0, \quad \operatorname{div}_h(\operatorname{grad}_h \phi) = \Delta \phi.$$

## 3. Parabolic Regular Equations of the Second Kind and IVP

Taking into consideration these equalities, one can obtain that the solutions of (3.4) are also solutions of the equation

$$\frac{\partial^2 v}{\partial x_0^2} + \frac{\partial^2 v}{\partial x_1^2} - \frac{\partial^2 v}{\partial x_2^2} = \frac{\partial v}{\partial x_3}. \qquad (3.6)$$

Now IVP will be considered. First note that for equation (1) Cauchy's problem with the condition

$$u(x, 0) = \phi(x), \quad x(x_0, \ldots, x_{n-1}) \in R^n, \quad x_n \equiv t > 0, \qquad (3.7)$$

is correctly posed and is represented by (3). But for the equation

$$\Delta u = -\frac{\partial u}{\partial t}, \quad x \in R^n, \quad t > 0, \qquad (3.8)$$

this problem is not correctly posed; a bounded solution does not exist. In fact, by the FIT of equation (3.8) with respect to the variables $x_0, \ldots, x_{n-1}$, one can obtain

$$\frac{d\widehat{u}}{dt} = |\xi|^2 \widehat{u},$$

which has no bounded solution for $t > 0$, $\xi \in R^n$. However, the parabolic equation of the second kind has quite different properties.

**Problem.** Find the regular solution of equation (3.1) for $x \in R^n$, $x_n = t > 0$, with the condition

$$u(x, 0) = \phi(x), \quad n \geq 2. \qquad (3.9)$$

**Solution.** By the FIT of equation (3.1) with respect to the variables $x_0, \ldots, x_{n-1}$, one can get

$$\frac{d\widehat{u}}{dt} = -(|\xi|^2 - \xi_{n-1}^2)\widehat{u}, \quad \xi(\xi_0, \ldots, \xi_{n-2}), \quad t > 0,$$
$$\widehat{u}(\xi, \xi_{n-1}, 0) = \widehat{\phi}(\xi, \xi_{n-1}).$$

Thus, the solution that vanishes at infinity can be represented in the form

$$\widehat{u} = \begin{cases} \widehat{\phi} e^{-(|\xi|^2 - \xi_{n-1}^2)t}, \\ \quad \text{when } |\xi|^2 > \xi_{n-1}^2, \\ 0, \\ \quad \text{when } |\xi|^2 \leq \xi_{n-1}^2. \end{cases} \qquad (3.10)$$

But if in place of equation (3.1), we consider the equation

$$\Delta u - \frac{\partial^2 u}{\partial x_{n-1}^2} = -\frac{\partial u}{\partial t}, \quad t > 0, \quad x \in R^n, \qquad (3.11)$$

with the condition (3.9), by the FIT one has

$$\widehat{u} = \begin{cases} 0, \\ \quad \text{when } |\xi|^2 \geq \xi_{n-1}^2, \\ \widehat{\phi} e^{(|\xi|^2 - \xi_{n-1}^2)t}, \\ \quad \text{when } |\xi|^2 < \xi_{n-1}^2. \end{cases} \qquad (3.12)$$

Solution (3.10) implies that the function $\phi(x)$ for equation (3.1) must have the property that its FIT is zero outside the domain $|\xi|^2 > \xi_{n-1}^2$. But (3.12) implies that for equation (3.11), the FIT of $\phi(x)$ must be zero outside of $|\xi|^2 < \xi_{n-1}^2$. Thus, for equations (3.1) and (3.11), the domain for the Cauchy problem must be correspondingly defined. In particular, in the case of (3.10), the domain $\Omega_1 : |\xi|^2 - \xi_{n-1}^2 \geq \varepsilon^2$ will be considered, and in the case (3.11), the domain $\Omega_2 : |\xi|^2 - \xi_{n-1}^2 \leq \varepsilon^2$. Then, using the inverse theorem, the solution of (3.9) can be represented in the form

$$u(x, t) = \frac{1}{(\sqrt{2\pi})^n} \int_{\Omega_1} \widehat{\phi}(\xi) e^{-(|\xi|^2 - \xi_{n-1}^2)t} e^{i \sum_0^{n-1} \xi_k x_k} d\xi_0 \cdots d\xi_{n-1}.$$

We have an analogous representation for equation (3.11) on $\Omega_2$. It is not possible, in general, to simplify the representations as explicit convolutions.

All the problems considered in the previous section can be considered for equation (7) in the space $R^0_{(n,n-1)}$ too. But because they can be reduced to the solution of Cauchy's problem of equation (3.1), their investigation will not be so evident as in $R^0_{(n)}$. Thus for equations in the space $R^0_{(n,n-1)}$, one has more complicated problems than in $R^0_{(n)}$, so for the classical heat equation which is connected to $R^0_{(n)}$, many problems can be solved explicitly.

It is clear that in $R^0_{(n,n-1)}$ polyheat equations corresponding to (3.1) can be considered too, but the solutions of IVP can not be represented in quadratures by evident formulae. That is why they are not considered here.

# 4  Elliptic-Parabolic, Hyperbolic-Parabolic and Elliptic-Hyperbolic-Parabolic Equations

Consider the equations

$$\overline{\partial}(\overline{\partial} u + P_n u) = 0, \tag{4.1}$$

$$\left(\overline{\partial} + e_{n-1}\frac{\partial}{\partial x_{n-1}}\right)(\overline{\partial} u + P_n u) = 0, \tag{4.2}$$

$$\overline{\partial}\left(\overline{\partial} + e_{n-1}\frac{\partial}{\partial x_{n-1}}\right)(\overline{\partial} u + P_n u) = 0, \tag{4.3}$$

where in (4.1)

$$\overline{\partial} = \sum_0^n \frac{\partial}{\partial x_k} e_k, \quad e_k^2 = -e_0, \quad k = 1, \ldots, n-1, \quad e_n^2 = 0, \tag{4.4}$$

and in (4.2) and (4.3)

$$\overline{\partial} = \sum_0^{n-2} \frac{\partial}{\partial x_k} e_k + \frac{\partial}{\partial x_n} e_n, \quad e_n^2 = 0,$$
$$e_k^2 = -e_0, \quad k = 1, \ldots, n-2, \quad e_{n-1}^2 = e_0. \tag{4.5}$$

Since $\partial\overline{\partial} = \Delta$, where $\Delta$ is the Laplace operator with respect to variables $x_0, \ldots, x_{n-1}$ in the case (4.4) and to variables $x_0, \ldots, x_{n-2}$ in the case (4.5), $P_n u$ is defined by (8). Then one can see that the solutions of (4.1), (4.2), (4.3) are also the solutions of the equations

$$\Delta\left(\Delta - \frac{\partial}{\partial t}\right) u(x, t) = 0, \quad x(x_0, \ldots, x_{n-1}), \quad x_n \equiv t > 0, \tag{4.6}$$

$$\left(\Delta - \frac{\partial^2}{\partial \tau^2}\right)\left(\Delta - \frac{\partial}{\partial t}\right) u(x, \tau, t) = 0, \quad x(x_0, \ldots, x_{n-2}), \quad x_{n-1} \equiv \tau, \quad x_n \equiv t > 0, \tag{4.7}$$

$$\Delta\left(\Delta - \frac{\partial^2}{\partial \tau^2}\right)\left(\Delta - \frac{\partial}{\partial t}\right) u(x, \tau, t) = 0, \tag{4.8}$$

respectively. Equations (4.6), (4.7), (4.8) are called the harmonic-heat, wave-heat and harmonic-wave-heat equations respectively.

First we will consider B&IVP for (4.6).

**Dirichlet–Cauchy and Neumann–Cauchy problems.** Find the regular solution of (4.6) for $x_{n-1} > 0$, $t > 0$, $(x_0, x_1, \ldots, x_{n-2}) \in R^{n-1}$, that vanishes at infinity by the conditions

$$u(x,0) = \varphi(x), \quad x_{n-1} > 0, \quad x(x_0, \ldots, x_{n-1}), \tag{4.9}$$
$$u(x,t) = \psi(x_0, x_1, \ldots, x_{n-2}, t), \quad x_{n-1} = 0, \quad t > 0, \tag{4.10}$$

or (4.9) and

$$\frac{\partial u}{\partial x_{n-1}} = \psi(x_0, \ldots, x_{n-2}, t), \quad x_{n-1} = 0, \quad t > 0. \tag{4.11}$$

**Solution.** Let

$$\Delta u(x,t) = F(x,t), \tag{4.12}$$
$$\Delta F - \frac{\partial F}{\partial t} = 0. \tag{4.13}$$

Then by force of (4.9), the unknown function $F(x,t)$ satisfies

$$F(x,0) = \Delta \varphi(x) \equiv f(x), \tag{4.14}$$

and one has for (4.13) Cauchy's IVP which is represented by (3). To define $u(x,t)$ we have the Dirichlet problem (4.10) or the Neumann problem (4.11) for equation (4.12), and the solutions are given in the first part. It is clear that all problems solved for harmonic functions in the first part can be solved correspondingly for equation (4.6).

**Cauchy's problem for equation (4.7).** Find the regular solutions of equation (4.7) for $t > 0$, $\tau > 0$, $x \in R^{n-1}$, by the conditions

$$u(x,0,t) = \varphi_1(x,t), \quad \frac{\partial u}{\partial \tau} = \varphi_2(x,t), \quad \tau = 0, \tag{4.15}$$
$$u(x,\tau,0) = \psi(x,\tau), \tag{4.16}$$

with the compatibility conditions $\varphi_1(x,0) = \psi(x)$.

**Solution.** Let

$$\Delta u - \frac{\partial u}{\partial t} = F(x,\tau,t), \tag{4.17}$$
$$\Delta F - \frac{\partial^2 F}{\partial \tau^2} = 0. \tag{4.18}$$

Then by force of (4.15), the unknown functions $F(x,\tau,t)$ satisfy

$$F(x, 0, t) = \Delta\varphi_1(x, t) - \frac{\partial\varphi_1(x, t)}{\partial t} \equiv f_1(x, t), \quad \tau = 0,$$
$$\frac{\partial F}{\partial \tau} = \Delta\varphi_2(x, t) - \frac{\partial\varphi_2(x, t)}{\partial t} \equiv f_2(x, t), \quad \tau = 0.$$
(4.19)

Thus for the wave equation (4.18), we have Cauchy's problem the solution of which is given in the first chapter of the second part. After defining $u(x, \tau, t)$, we have Cauchy's problem for the nonhomogeneous heat equation (4.17) with the condition (4.16). The solution is represented in the form

$$u(x, \tau, t) = u_1(x, \tau, t) + u_2(x, \tau, t),$$

where $u_1(x, \tau, t)$ is the solution of the homogeneous heat equation with the condition $u_1(x, \tau, 0) = \psi_1(x, \tau)$ and $u_2(x, \tau, t)$ is the solution of (4.17) with the condition $u_2(x, \tau, 0) = 0$. Thus using (3) and (5), the solution can be represented in quadratures.

It is obvious that one can consider the heat-Klein–Gordon equation

$$\left(\Delta - \frac{\partial}{\partial t}\right)\left(\Delta - k^2 - \frac{\partial^2}{\partial \tau^2}\right) u(x, \tau, t) = 0,$$

with the conditions (4.15), (4.16), and the solution is represented in quadratures.

In the same way one can consider the Helmholtz-heat equation

$$(\Delta - k^2)\left(\Delta - \frac{\partial}{\partial t}\right) u(x, t) = 0,$$

with the conditions (4.9), (4.10) or (4.9), (4.11). The solutions can be represented in quadratures too.

Now consider the problem for the harmonic-wave-heat equation (4.8) for $t > 0$, $\tau > 0$, $x_{n-2} > 0$ with the conditions

$$u(x, 0, t) = f_1(x, t), \quad \frac{\partial u}{\partial \tau} = f_2(x, t), \quad \tau = 0, \quad (4.20)$$
$$u(x, \tau, 0) = \varphi(x, \tau), \quad (4.21)$$
$$u(x, \tau, t) = \psi(x_0, \ldots, x_{n-3}, \tau, t), \quad x_{n-2} = 0, \quad (4.22)$$

with the corresponding compatibility conditions.

**Solution.** Let

$$\Delta u = F(x, \tau, t), \quad (4.23)$$

168   IV. Parabolic and Pluriparabolic Equations in Clifford Analysis

$$\left(\Delta - \frac{\partial^2}{\partial \tau^2}\right)\left(\Delta - \frac{\partial}{\partial t}\right) F = 0. \tag{4.24}$$

Then by force of (4.20), (4.21), $F$ satisfies conditions like (4.15), (4.16), i.e., $F$, as solution of (4.24), is constructed effectively. Hence $u$, as a solution of (4.23) by the condition (4.22) or the condition of the Neumann problem, can be represented in quadratures too.

Note that if we consider the problem for the equation

$$\left(\Delta - \frac{\partial^2}{\partial t^2}\right)\left(\Delta - \frac{\partial}{\partial t}\right) u(x, t) = 0, \quad t > 0, \quad x(x_0, \ldots, x_{n-1}), \tag{4.25}$$

with the conditions

$$u(x, 0) = f_1(x), \quad \frac{\partial u}{\partial t} = f_2(x), \quad \frac{\partial^2 u}{\partial t^2} = f_3(x), \quad t = 0, \tag{4.26}$$
$$u(x, t) = f(x_0, \ldots, x_{n-2}, t), \quad x_{n-1} = 0,$$

then the solution is defined in the same way as for equation (4.7) with conditions (4.15), (4.16).

We think that equations (4.7), (4.8) are more interesting than (4.25) firstly because they are related to equations (4.2), (4.3) (i.e., they are suggested by Clifford analysis) and secondly because it is natural that the time in the wave processes and the time in the heat processes would be different. That is why equations (4.1), (4.2), (4.3), (4.6), (4.7), (4.8) probably have important applications in physics.

Now, to formulate the B&IVP for equations (4.1), (4.2), (4.3), all those conditions that are considered for the multiplier operators must be given. Thus using the solutions of each problem, one can obtain corresponding solutions in quadratures.

B&IVP for the nonhomogeneous equations that correspond to the above considered homogeneous equations can be solved too. It is obvious in this case that the boundary-initial conditions can be supposed homogeneous. We will consider only one of them. Others can be solved in the same way.

**Problem.** Find the solution of the equation

$$\Delta\left(\Delta - \frac{\partial^2}{\partial \tau^2}\right)\left(\Delta - \frac{\partial}{\partial t}\right) u(x, \tau, t) = F(x, \tau, t), \tag{4.27}$$
$$x(x_0, \ldots, x_{n-2}) \in R^{n-1}, \quad \tau > 0, \quad t > 0, \quad x_{n-2} > 0,$$

that vanishes at infinity by the conditions

$$u(x, 0, t) = 0, \quad \frac{\partial u}{\partial \tau} = 0, \quad \tau = 0, \tag{4.28}$$
$$u(x, \tau, 0) = 0, \tag{4.29}$$

$$u(x, \tau, t) = 0, \quad x_{n-2} = 0. \tag{4.30}$$

**Solution.** Let

$$\begin{aligned}
\Delta u &= F_1(x, \tau, t), \\
\Delta F_1 - \frac{\partial^2 F_1}{\partial \tau^2} &= F_2(x, \tau, t), \\
\Delta F_2 - \frac{\partial F_2}{\partial \tau} &= F(x, \tau, t).
\end{aligned} \tag{4.31}$$

Then by force of (4.28), (4.29), for $F_1$, $F_2$ one has Cauchy homogeneous conditions for nonhomogeneous wave and heat equations. Thus they are defined in quadratures. Hence $u(x, \tau, t)$ is defined as the solution of (4.31) with condition (4.30).

It seems to me that these equations are beautiful. That is why, as Paul Dirac said about beautiful formulae, their success in applications is ensured.

# Epilogue

Paul Dirac—the famous English physicist and mathematician—wrote in his excellent essay "Recollections of an Exciting Era," "I always aspired to obtain the beautiful formulas." The mathematical beauty of equations describing natural law was a symbol of belief for him; he considered it to be the basis of his important success. Dirac used to write "Nature" with a capital "N."

Indeed, in beautiful formulae, theory is truth. Fifty years before publishing that essay, Dirac made one of the greatest inventions of twentieth century. In a theoretical way, he discovered a new particle in the nucleus called the positron using only simple mathematics. Two years later, the existence of this particle was proved experimentally. For that remarkable discovery, Dirac was awarded the Nobel Prize in 1933.

But is it easy to obtain beautiful formulae? No, on no account. We mathematicians write and delete, write and delete, many times. We are sometimes disappointed and sometimes delighted. When at last beautiful formulae are found, and especially if they are found in a simple way, we are happy.

In this book, we tried to consider new or known equations and problems the solutions of which can be represented in quadratures in beautiful forms. And this is true happiness and creative joy for us.

# References

(This list does not claim to be complete; mainly the monographs are cited.)

[Be] Begehr, H., *Complex Analytic Methods for PDE*, World Scientific, Singapore, London, 1994.

[BG] Begehr, H., and Gilbert, R. *Transformations and Kernel Functions*, vol. 1, Longman Scientific and Technical, Harlow, 1992.

[Bi] Bitsadze, A., *Some Classes of PDE*, Nauka, Moscow, 1981 (in Russian).

[BDS] Brackx, F., Delanghe, R., and Sommen, F., *Clifford Analysis*, Pitman, London, 1982.

[CH] Courant, R. and Hilbert, D., *Methods of Mathematical Physics*, vol. II, Interscience Publishers, New York, 1962.

[DSS] Delanghe, R., Sommen, F., and Souĉek, V., *Clifford Algebra and Spinor Valued Functions*, Kluwer Academic, Dordrecht, 1992.

[Di] Dirac, P., *Recollections of an Exciting Era*, Proceedings of the International School of Physics "Enrico Fermi," Academic Press, 1977, pp. 109–146.

[Ga] Gakhov, F., *Boundary Value Problems*, Pergamon Press, Oxford, 1966.

[GC] Gakhov, F, and Cherski, J., *Equations of Convolution Type*, Nauka, Moscow, 1978 (in Russian).

# References

[GS] Gurlebeck, K. and Sprossig, W., *Quaternionic Analysis and Elliptic Boundary Value Problems*, Akademie-Verlag, Berlin, 1989.

[Ha] Habetha, K., Function Theory in Algebras. *Complex Analysis. Trends and Applications*, ed. E. Lanckau and W. Tutschke, Akademie-Verlag, Berlin, 1985, pp. 225–237.

[He] Hestenes, D., *Space-Time Algebra*, Gordin and Breach, New York, 1966.

[KS] Kravchenko, V. and Schapiro, M., *Integral Representations for Spacial Models of Mathematical Physics*, Longman, Edinburg, 1996.

[Mu1] Muskelishvili, N., *Singular Integral Equations*, Noordhoff, Groningen, 1953.

[Mu2] Muskhelishvili, N., *Some Basic Problems of the Mathematical Theory of Elasticity*, Nauka, Moscow, 1966 (in Russian).

[N] Noble, B,, *Methods Based on the Wiener–Hopf Technique for the Solution of PDE*, Pergamon-Press, London, 1958.

[Ob1] Obolashvili, E., *PDE in Clifford Analysis*, Longman, Edinburgh, United Kingdom, 1998.

[Ob2] Obolashvili, E., Beltrami Equations and Generalizations in Clifford Analysis, *Applicable Analysis*, vol. 73(1-2), 1999, pp. 167–185.

[Ob3] Obolashvili, E., Some PDE in Clifford Analysis, *Complex Methods for PDE*, ed. H. Begehr, O. Celebi, W. Tutschke. Kluwer Academic, 1999, pp. 245–261.

[Ob4] Obolashvili, E., *Fourier Integral Transformation and Its Applications in Elasticity Theory*, Metsniereba Publishing House, Tbilisi, 1979 (in Russian).

[Ob5] Obolashvili, E., Some PDE in Clifford Analysis, *Banach Center Publications*, vol. 37, 1996, pp. 173–176.

[Ob6] Obolashvili, E., Some PDE in Clifford Analysis, *Advances in Geometric Analysis and Continuum Mechanics*, International Press, Cambridge, 1995, pp. 232–239.

[Ri] Riesz, M., Clifford Numbers and Spinors, *Lecture Series* 38, Maryland, 1958.

[Ry] Ryan, J., *Clifford Algebras in Clifford Analysis and Related Topics*, CRC Press, New York, 1996.

[SW] Stain, E. and Weiss, G., *Introduction to Fourier Analysis on Euclidean Spaces*, Princeton University Press, Princeton, New Jersey, 1971.

[Ti] Titchmarsh, E., *Introduction to the Theory of Fourier Integrals*, Clarendon Press, Oxford, 1961.

[Tu] Tutschke, W., *Solutions of Initial Value Problems in Classes of Generalized Analytic Functions*, Teubner, Leipzig, 1989.

[TS] Tychonoff, A. N. and Samarski, A. A., *Differentialgleichungen Der Mathematischen Physik*, Deutscher Werlag, Berlin, 1959.

[Ve] Vekua I. N., *Generalized Analytic Functions*, Pergamon Press, Oxford, 1962.

# Index

Bessel functions, 72

Cauchy kernel, 92
Cauchy problem, 125, 130, 133, 135, 139, 151, 156, 161
Cauchy type, 4
Cauchy–Riemann iterated operator, 44
Cauchy–Riemann operator, 4
compound BVP, 6, 101

Dini integral, 9
Dirichlet problem, 6, 10, 78, 83
Dirichlet–Cauchy problem, 141
dual integral equation, 55

extension theorem, 118

Fejér kernels, 70
Fourier integral transformation, 67
fundamental solution, 92, 93

Gauss formula, 77
Generalized Cauchy integral formula, 95
generalized Cauchy–Riemann system, 14

Goursat problem, 145
Green formula, 78, 83
Green functions, 4, 95

Hankel function, 73, 76, 92
harmonic-Helmholtz equation, 109
heat equation, 151
Helmholtz equation, 83
Helmholtz-heat equation, 167
Hilbert BVP, 5, 92, 99

Keldish–Sedov mixed problem, 11
Klein–Gordon equation, 125
Kolosov–Muskhelishvili representation, 15, 46

Liouville theorem, 154
Lorentz transformation, 143

Moisil–Theodorescu system, 88, 90, 117
Moivre formula, 117

Neumann problem, 8, 78–81, 83
Neumann–Cauchy problem, 142, 165

piecewise holomorphic function, 5, 12

plane crack, 106, 108
Plemelj–Privalov theorem, 4
Plemelj–Sokhotzki formulae, 3, 98
Poisson formula, 8, 86

Riemann–Hilbert BVP, 6, 101
Riemann–Schwartz principle of reflection, 9, 10, 25, 26, 81, 119
Riesz system, 91
Riquie problem, 15, 49, 51, 109

Schwartz formula, 7
Sommerfeld formula, kernel, 75, 76

wave equation, 125
Wiener–Hopf integral equation, 56